The
Turfgrass
Disease
Handbook

Houston B. Couch

Professor of Plant Pathology
Virginia Polytechnic Institute
and State University

KRIEGER PUBLISHING COMPANY
MALABAR, FLORIDA
2000

Original Edition 2000

Printed and Published by
KRIEGER PUBLISHING COMPANY
KRIEGER DRIVE
MALABAR, FLORIDA 32950

Library of Congress Cataloging-in-Publication Data

Couch, Houston B.
    The turfgrass disease handbook / Houston B. Couch.
        p.   cm.
    Includes bibliographical references.
    ISBN 1-57524-076-9 (cloth : alk. paper)
        1. Turfgrasses—Diseases and pests—Handbooks, manuals, etc.
    2. Turf management—Handbooks, manuals, etc. I. Title.
    SB608.T87 C69 2000
    635.9'6493—dc21
                                                                99-053330

10 9 8 7 6 5 4 3 2

To my grandchildren, Jonathan, Jessica, Joseph,
Sean, Joy, Daniel, Stephen, Joshua, and Rebecca,
for the wonderful dimension of happiness
they have added to my life.

Contents

Contents

Preface

The purpose of this book is to provide a concise, up-to-date source of information on the nature, diagnosis, and control of the major diseases of cool season and warm season golf, landscape, recreation, and sports turfgrasses. The material on each disease is presented through a series of sub topics: (1) The Name of the Pathogen, (2) Grasses Affected, (3) Season of Occurrence, (4) Symptoms and Signs, (5) Conditions Favoring Disease Development, and (6) Control. The control sections for many of the diseases are divided into three sections: (1) Cultural Practices, (2) Use of Resistant Grasses, and (3) Chemical Control. This format takes the reader sequentially through every aspect of disease diagnosis and control.

Specific features include color photographs showing disease symptoms, photomicrographs of pathogens, descriptions of on-site diagnostic procedures, and strategies for integrated disease control through the use of chemical pesticides in conjunction with management practices. The information on the nature of each disease gives the background needed to predict disease outbreaks in time to initiate preventive control measures, and in the event of severe occurrences of disease, to assess the potential for plant recovery.

Information is also provided on the impact of environmental conditions and cultural practices on the health of the turfgrass plant and how such factors as air and soil temperature extremes, mowing practices, fertilization and soil pH, watering programs, and thatch accumulation interact to either increase or decrease the turfgrass plant's resistance to pathogenic microorganisms. Specific chemical pesticides, listed by their coined names, are suggested for the control of each disease. A representative trade name, topical mode of action, and the manufacturer's address and telephone number for each pesticide is given in Appendix Table I.

I am very thankful to Dr. Bruce Martin of Clemson University, Dr. Philip Colbaugh of Texas A & M University, and Dr. Nick Christians of Iowa State University for reviewing the manuscript and making helpful suggestions. Appreciation also goes to Krieger Publishing Company of Malabar, Florida, for permission to reprint many of the illustrations from the third edition of *Diseases of Turfgrasses* (1995) by Houston B. Couch.

Houston B. Couch
Blacksburg, Virginia
September 1999

1

Causes of Turfgrass Diseases

The causes of turfgrass diseases have their origins in the plant's physical and biological environments. Those from the physical environment are referred to as **abiotic causal factors.** The individual abiotic factors fit into one of two groups: (1) climatic conditions that produce plant stress—such as air and soil temperature extremes, and wind desiccation of leaves, and (2) improper management practices—including extreme cutting heights, misapplication of pesticides, poorly timed irrigation schedules, and inappropriate root zone management resulting in either root dysfunction or death.

Causes that have their origins in the plant's immediate biological environment are called **biotic disease inciting entities.** The disease inciting agents in this category include any life form that is known to have a negative impact on the capacity of the plant's cells to function normally. Examples of members of this group are fungi, nematodes, viruses, bacteria and algae.

Correct identification of the causal factors is an essential component of a successful disease control program. This section of the handbook identifies the primary biotic and abiotic incitants of turfgrass diseases and describes how each impacts on the well-being of the plants.

Abiotic Factors That Cause Turfgrass Diseases

Examples of abiotic factors that often cause turfgrass to become diseased are air pollutants, excesses and/or deficiencies in plant nutrient levels, soil moisture stress, anaerobic soil conditions, excessively close mowing over an extended period of time, lightning strikes, extremes in air temperatures, pesticide injury, and wind desiccation of the leaves.

The symptoms of gaseous air pollutant damage to individual turfgrass leaves range from uniform yellowing, to tip die-back, to distinctive dark brown spots and streaks. In temperate climates, turfgrass going into the winter season at low soil moisture levels is highly vulnerable to plant death caused by wind desiccation. This problem is particularly severe on turf located at high elevations in the landscape—such as the crest of hills, ridges, and the tops of mounds. Also, turfgrass growing in these locations is more vulnerable to frost damage.

Lightning strikes on turf will cause streaks of brown grass of different lengths that radiate outward in starlike fashion from the point of impact. Where extremes in plant nutrients levels are concerned, nitrogen excess will at first cause the leaves to become dark green, then turn yellow, and finally brown. Nitrogen deficiency will bring about slow plant growth and leaf yellowing. Excessively close mowing of the turf will cause a significant decrease in carbohydrate reserves. During times of heat stress, these plants are highly vulnerable to root deterioration and yellowing and death of leaves.

Poor surface drainage or low water infiltration rates can cause soil pores to remain filled with water for long periods of time. This condition leads to a depletion of oxygen in the root zone. When the oxygen level in the soil is deficient, the availability of nutrients changes, and the ability of the turfgrass root system to absorb water is hampered. When this happens, deficiency symptoms of the major nutrient elements develop in the leaves and the rate of plant growth decreases. Unless the infiltration rate and surface drainage problems are corrected, plant death will occur.

Biotic Entities That Cause Turfgrass Diseases

Biotic entities that cause diseases of turfgrasses include viruses, mollicutes, bacteria, nematodes and fungi. Viruses are pathogenic, submicroscopic, infectious particles that multiply only within living host cells. Seven diseases of turfgrasses are known to be caused by viruses. Mollicutes are the smallest known free-living organisms. Two types of mollicutes, spiroplasma and mycoplasma, are known to be pathogenic to ryegrass and bermudagrass. Bacteria are distinguished from mollicutes by their larger size and the development of a rigid cell wall with outer and

Figure 1-1. Photomicrograph of developmental stages of the spiral nematode, a species that parasitizes turfgrass roots. From left to right: egg, newly hatched juvenile, first, second and third stage juveniles, and mature female. *Courtesy A. H. Golden.*

inner membranes. A species of bacterium is known to cause wilt and leaf death of bentgrasses, ryegrasses, fine and tall fescues, Kentucky bluegrass, and annual bluegrass.

The nematodes that parasitize turfgrass roots are very small, averaging 1/25 inch (0.1 mm) in length. They are wormlike in appearance but are very distinct taxonomically from segmented worms (Figure 1-1). Leaf and soil inhabiting species of parasitic nematodes can cause extensive damage to established turf. Nematodes feed on leaf and root cells by means of a syringelike structure known as a **stylet** (Figure 1-2). The stylet serves to inject digestive enzymes into the cell and then withdraw the partially digested contents. In order for root-feeding nematodes to have an adverse effect on overall turf quality, however, they must be able to produce very high population levels within a relatively short period of time.

Fungi are members of the kingdom Myceteae. They are nongreen organisms that do not

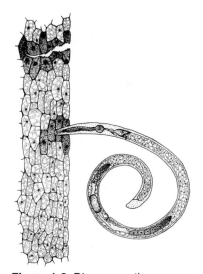

Figure 1-2. Diagrammatic representation of an ectoparasitic nematode feeding on a grass root. Note position of head and stylet. *From Couch, 1995.*

produce true seeds, and their bodies lack true roots, stems and leaves. Since fungi do not produce chlorophyll, they must obtain their nutrition as saprophytes growing on dead organic matter or by parasitizing other life forms. More than 8,000 species of fungi have been found to be pathogenic to higher plants. Of this group, 127 species are known to cause foliar, crown, and root diseases of turfgrasses. In most of the regions of the world, there is at least one fungus-incited disease for each turfgrass species for each season of the year.

The body of a fungus is a very simple structure. It is either made up of a single cell or a series of cells arranged end-to-end in tubelike strands. An individual strand is called a **hypha** (pl. = hyphae). The hyphae of individual species may vary from .00001968 inch (0.5 μm) to .003937 inch (100 μm) in diameter. An aggregate of a group of hyphae is called a **mycelium** (pl. = mycelia). Reproduction of many fungi is accomplished by **spores.** Spores are analogous to seeds in plants. However, they differ from true seeds in that they do not contain embryonic tissue. Certain fungi, such as the mushrooms and puffballs that cause fairy rings and localized dry spots in established turf, produce their spores on or within visible fruiting bodies which are formed by the aggregation of numerous hyphae. Most of the fungus species that are pathogenic to turfgrasses, however, do not produce highly conspicuous spore-bearing structures. Instead, the spores are either developed in very small specialized bodies, or are scattered openly over the surface of the mycelium (Figure 1-3).

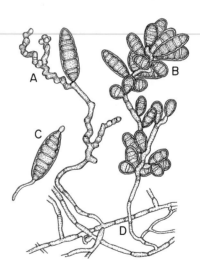

Figure 1-3. Growth pattern of the fungus that causes Helminthosporium leaf spot. A) Spore-bearing structure showing former places of attachment of spores. B) Cluster of spores still attached to the hypha. C) Spore germinating by means of germ tubes from the two terminal cells. D) Hyphal strands. *From Couch, 1995.*

Some of the fungi that are pathogenic to turfgrasses invade the leaves through open stomates or wounds such as mower cuts; however, the majority infect the leaves by directly penetrating the surfaces of intact epidermal cells. Their spores germinate to produce

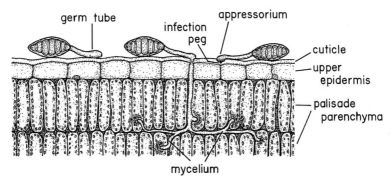

Figure 1-4. Illustration of how the fungus that causes Helminthosporium leaf spot penetrates the epidermis of a turfgrass leaf and then begins to parasitize the underlying cells. *From Couch, 1995.*

what are known as **germ tubes,** which then elongate into hyphae. As the hyphae grow over the surface of the leaf, they either aggregate into small mats of mycelium known as **infection cushions** which serve to facilitate penetration of the cuticle or individually develop or develop specialized penetration structures at their tips known as **appressoria.** The appressoria attach to the cuticle and produce infection pegs at the site of contact. By a combination of physical pressure and the secretion of enzymes that degrade cutin, the infection pegs penetrate the cuticle. A hyphal strand

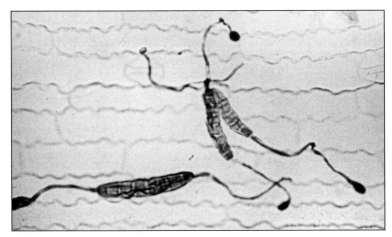

Figure 1-5. Photomicrograph of direct penetration of leaf surface by means of appressoria produced by germinating *Bipolaris* spores. *Courtesy Philip Larsen.*

then develops from the terminal portion of each infection peg. Then, through a combination of physical pressure and secretion of enzymes that degrade pectin and cellulose, these hyphae invade and parasitize the cells of the underlying tissues (Figure 1-4; Figure 1-5).

Cenchrus echinatus
(southern sandbur)

2

Diagnostic Procedures

Introduction

Accurate and early diagnosis is an essential component of a successful turfgrass disease control program. Damaging turfgrass diseases are often preceded by relatively minor and somewhat isolated disease outbreaks in restricted areas. For this reason, regardless of severity, every effort should be made to determine the cause of each episode of disease and preventive measures taken to prevent its further spread.

The Complexities of Diagnosis

Diagnosis of turfgrasss diseases is not always a simple task. Successful on-site diagnosis requires that one have a working knowledge of the comparative symptom patterns of all diseases that could be occurring on the turf species under consideration during the season at hand. This includes not only being familiar with the so-called typical symptoms, but also having an understanding of how certain primary and secondary symptoms associated with each disease vary (a) with changes in plant growth management practices, (b) with different climatic conditions, and (c) among grass species.

For example, where cutting height is concerned when creeping bentgrass (*Agrostis palustris*) is cut at 1/2 inch (1.3 cm), Pythium blight and Helminthosporium leaf spot on this species are easily distinguished from each other. Pythium blight develops as leaf withering and loss of leaf firmness, while in contrast the primary symptom pattern of Helminthosporium leaf spot is clearly delineated leaf lesions with the leaves retaining their shape. On the other hand, when bentgrass is managed under the ultralow mowing heights used for golf greens and bowling greens, the overall symptom pattern of both of these diseases is very similar. They both develop as irregularly shaped areas of turf with a smoky-blue cast that is soon followed by a total blighting and matting together of the leaves.

The various symptom patterns of Fusarium patch illustrate how field diagnostic features of turfgrass diseases can vary with climatic conditions. During cold, wet weather the patches that develop on bentgrass (*Agrostis* spp.) golf greens are tan with white to pinkish mycelial growth at the margins giving them a distinctive pinkish cast. In contrast, under snow cover the patches produced by the disease have a bleached appearance, and the abundant mycelial growth and pink borders are absent.

Rhizoctonia blight serves as an illustration of how symptoms can vary among certain turfgrass species. On zoysiagrass (*Zoysia japonica* and *Zoysia tenuifolia*) this disease is characterized by the development of large patches of yellow turf. There are no lesions on the leaf blades but the bases of the leaf sheaths are colonized by the pathogen, causing a distinctive dark brown, soft rot of the tissue, enabling the diseased leaves to be easily pulled loose from the crowns. In contrast, on tall fescue (*Festuca arundinacea*), in overall view, the patches of affected turf are brown. Distinctive light tan lesions with reddish-brown borders develop on the leaf blades. Also, although colonization of the leaf sheaths by the Rhizoctonia blight pathogen does occur, the necrotic tissue retains a firm texture.

Diagnosis can be further complicated by the fact that during the season at hand two or more diseases that have similar general symptom patterns can develop on the same stand of turfgrass. For instance, Fusarium blight, summer patch, necrotic ring spot, Pythium blight, and Rhizoctonia blight are patch diseases that may affect the same cool season turfgrass species during spring and summer months. Also, each disease often develops the classic "frogeye" symptom pattern of distinctive circular patches of blighted grass with center areas of green, apparently healthy plants.

Procedures for On-Site Diagnosis
When examining the diseased area, one should locate the sections in the turf where there is some degree of demarkation between zones of diseased and healthy grass. These transition areas between diseased and healthy grass will provide the greatest range of early and advanced symptoms of the disease at hand.

Since two or more diseases often develop simultaneously in the same turf, areas within the affected site should also be checked for possible differences in overall symptom patterns. Individual plants from each location should then be inspected. These examinations should involve a thorough check of the state of health of all plant parts—leaves, leaf sheaths, crowns, stolons, rhizomes, and roots. Material from healthy plants should be compared with similar organs from plants in early, intermediate, and advanced stages of disease development. The following distinguishing features of the diseased tissue should be carefully noted: (a) the number, size, and color patterns of leaf lesions that might be present, (b) the general condition of the crowns and crown buds, and (c) the color, density, and depth of the feeder root systems. In addition to determining if more than one disease is involved, this procedure also indicates the likelihood of plant recovery if or when a pesticide program is initiated, or if weather favorable for turf growth is anticipated.

These observations should then be compared with photographs and descriptions in books and field manuals on grass diseases for the diseases known to occur on the grass species in question at the time of year the problem has developed. Also, in cases of first diagnosis of a specific disease, it is well to seek confirmation by specialists from private, state, or provincial diagnostic services or by local, experienced turfgrass management specialists.

Using the Services of Plant Disease Diagnostic Laboratories
The complexities of on-site diagnosis make it necessary at times to have a laboratory workup performed on samples of diseased turf to check for

the presence of known biotic pathogens. States and provinces operate plant disease diagnostic laboratories either through their public universities or agricultural advisory services. Also, many areas have private laboratories that perform turfgrass disease diagnoses. Appendix Table II contains a list of local sources of information on diagnosis and control of turfgrass diseases in the United States and Canada.

Samples collected for laboratory workups should be taken from within the areas of most severe damage and from adjacent locations with both healthy and diseased plants. A knife, small shovel or golf green cup cutter should be used to remove a section of sod 4 to 6 inches (10–15 cm) in diameter and to the depth of the root system.

Soil samples taken specifically for nematode assay should be collected from areas immediately outside the section of turf showing stress symptoms. The reason for this is that most plant parasitic nematodes are free-living obligate parasites; therefore, the highest population levels will be in soil that contains actively growing turfgrass roots. The sampling procedure should be performed after the soil temperature at the 2 inch (5 cm) depth reaches 50 °F (10 °C) or higher.

Using a standard 1 inch (2.5 cm) diameter soil tube, take 15 to 30 cores per 500 to 1,000 square feet (46–93 m^2). Samples should be taken to a depth of 4 inches (10–15 cm), not including the thatch. The individual samples from the site in question can be mixed together. The total volume of soil from each of these locations should be approximately 1 pint (500 ml), which is approximately 20 soil cores taken to a 4-inch depth.

If the samples are to be mailed or shipped immediately to the diagnostic laboratory by an express delivery service, they should be sealed in plastic bags and packaged very tightly to reduce the possibility of being broken up while in route. Information accompanying the samples should include (a) name, address, and telephone number of the sender, (b) site location of each sample, (c) species and varieties of grass comprising the sample, (d) overall size of the areas affected and description of the symptoms, (e) weather conditions for one week preceding the outbreak of the disease, including daytime and nighttime air temperatures, (f) fertilization dates and rates, and (g) a list of all fungicides, growth regulators and herbicides used for the previous 12 months.

3

Patch Diseases

Introduction

The "patch" disease symptom pattern in turfgrasses is one in which the overall appearance of the turf is characterized by a blighting of the majority of the leaves of the plants in individual sections of otherwise green sod. This group of diseases makes up a large and important segment of the major diseases of turfgrasses. Throughout the world there is at least one patch disease for every cultivated warm and cool-season turfgrass species for every season of the year.

Patch diseases are sometimes difficult to diagnose in the field because certain of the more striking symptoms associated with patch diseases can also be incited by a variety of causes other than the pathogenic microorganisms. Plant stress brought on by extremely high or extremely low air temperatures, low soil moisture content, water saturated soil, improper mowing, or improper fertilization practices can also bring about a browning of turfgrass in irregularly shaped patches.

Field diagnosis of certain patch diseases can be further complicated by the fact that they often have certain symptoms in common. For example, the frogeye pattern (circular patches of blighted grass with center tufts of green, apparently healthy plants) is common to several diseases in this group. This means that accurate diagnosis of patch diseases caused by pathogenic microorganisms requires the diagnostician to be familiar with the full range of primary and secondary symptoms common to these diseases as a whole and then be able to identify the specific features that are unique to each disease.

The various patch diseases described in this section of the handbook are grouped according to the season of the year in which they are usually most severe: Spring and Fall Patch Diseases, Summer Patch Diseases, and Winter Patch Diseases.

SPRING AND FALL PATCH DISEASES
Necrotic Ring Spot

- **Pathogen:**

Leptosphaeria korrae

- **Grasses Affected:**

Annual bluegrass (*Poa annua*), creeping bentgrass (*Agrostis palustris*), Kentucky bluegrass (*Poa pratensis*), Chewing's fescue (*Festuca rubra* var. *commutata*), red fescue (*Festuca rubra*), perennial ryegrass (*Lolium perenne*)

- **Season of Occurrence:**

Late winter, spring, and fall

- **Symptoms and Signs:**

In the early stages of disease development, necrotic ring spot is seen as irregular patches of grass that have a general appearance of drought injury. The plants are often stunted or discolored and turn various shades of red, yellow, or tan. These areas become dull tan to brown as the disease progresses (Plate 3-1).

The shapes of the individual patches of dead grass are usually more or less circular in outline. They may range in size from 2–3 inches (5–8 cm) to 2–6 feet (0.6–2 m) in diameter. At first, leaf death is usually uni-

Plate 3-1. Overview of necrotic ring spot on Kentucky bluegrass lawn.

form throughout the affected area. However, as the disease progresses, many of the patches may develop center tufts of apparently disease-free grass. This combination produces a distinctive frogeye effect. During weather conditions that are particularly favorable for outbreaks of necrotic ring spot, reddish-brown borders may develop between the patches of dead plants and the adjacent healthy grass (Plate 3-2; Plate 3-3). The diseased plants can be easily lifted from the soil.

- **Conditions Favoring Disease Development:**

Development of necrotic ring spot is generally most active during the cool, wet weather of spring and fall. During April and

Plate 3-2. Mycelial mass of necrotic ring spot pathogen on leaf sheath of Kentucky bluegrass. *Courtesy Gale Worf.*

Plate 3-3. Necrotic ring spot of Kentucky bluegrass showing leaf death at advancing margins of patches. *Courtesy Noel Jackson.*

May, heavy outbreaks of the disease have been noted after prolonged periods of rainfall. Necrotic ring spot is usually more destructive in stands of turfgrass under high nitrogen fertilization. The severity of the disease is not affected by soil pH's in the 5.0–8.0 range.

Outbreaks of necrotic ring spot are usually most prominent in three to four year old turfs. Spread of the pathogen to new sites is accomplished primarily by the transport of infested soil and diseased crown and root tissue on coring, vertical mowing, and power raking equipment.

• **Control:**

1. Cultural Practices—Necrotic ring spot is most severe on Kentucky bluegrass, annual bluegrass, and creeping red fescue. During the months of spring and early fall, golf greens or bowling greens with high populations of annual bluegrass should be carefully monitored for outbreaks of this disease. Management practices that promote deep rooting of the turfgrass during periods of new root growth (spring and fall), will aid materially in reducing the severity of outbreaks of necrotic ring spot.

2. Use of Fungicides—Azoxystrobin, fenarimol, thiophanate methyl, myclobutanil, and propiconazole have been reported to control necrotic ring spot. See Appendix Table I for a listing of representative trade names and manufacturers for each of these fungicides.

In order to obtain maximum disease control with these materials, while the leaves are still wet with the spray the treated areas should be watered to a soil penetration depth of 1 inch (2.5 cm). Spiking or coring prior to treatment facilitates movement of the fungicide into the root zone. Also, the fungicide will move more readily through soils that are initially moist than through soils that are dry at the time of application. Timing of the initial fungicide application is important to maximum control of the disease. Treatments should begin when the soil temperature at the 3 inch (7.5 cm) depth reaches 60 °F (16 °C) and continue at 30 day intervals for as long as weather conditions are favorable for disease development.

Take-all Patch

- **Pathogen:**
Gaeumannomyces graminis var. *avenae* (syn. *Ophiobolus graminis*) (Plate 3-4).
- **Grasses Affected:**
Bentgrasses (*Agrostis* spp.), annual bluegrass (*Poa annua*), Kentucky bluegrass (*Poa pratensis*), red fescue (*Festuca rubra*), tall fescue (*Festuca arundinacea*), perennial ryegrass (*Lolium perenne*), rough stem bluegrass (*Poa trivialis*)
- **Season of Occurrence:**
Late winter, spring, and fall
- **Symptoms and Signs:**
The bentgrasses (*Agrostis* spp.) are highly susceptible to take-all patch. This disease is first seen as depressed, circular areas of blighted turfgrass 2–3 inches (5–8 cm) in diameter. At this stage of development, the color of the patches may range from dull to bright reddish bronze. During the growing season, the color fades from dull brown to tan (Plate 3-5).

Affected individual areas may eventually reach a diameter of 2 feet (0.6 m) or larger. The centers of the patches frequently fill in with resistant species, thus creating a frogeye appearance of a green patch of grass, surrounded by a ring of bronze-colored plants. Often the individual patches will coalesce to form large irregularly-shaped areas of dead turf (Plate 3-6; Plate 3-7).

Roots colonized by the take-all pathogen first develop internal dark brown necrotic streaks. With the advent of warm, dry weather, they then turn dark brown to black and become brittle. At this time, the plants can easily be pulled loose from the soil. Individual strands of dark brown runner hyphae develop on

Plate 3-4. Fruiting structure (perithecia) of take-all pathogen on stem of *Poa trivialis.*

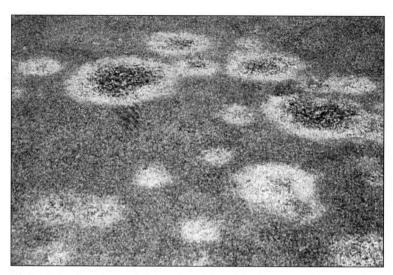

Plate 3-5. Overview of take-all patch of bentgrass. *Courtesy Gary Chastagner.*

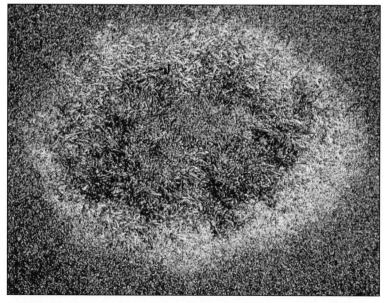

Plate 3-6. Take-all patch of creeping bentgrass showing weed fill-in growth in center of patch. *Courtesy Gary Chastagner.*

Plate 3-7. Take-all patch of creeping bentgrass showing characteristic bronze coloration at margins of patches when pathogen is active.

the surfaces of diseased rhizomes, stolons, and roots, and on the culms under the basal leaf sheaths. These can be seen with the aid of a wide field microscope. Also, in late fall, very small, dark brown to black, flask-shaped structures (perithecia) may develop on the culms of affected plants.

•   **Conditions Favoring Disease Development:**
Root infection by the take-all pathogen is favored by moist soil conditions and soil temperatures in the 50–65 °F (10–19 °C) range. The disease is initiated during the cool, wet months of early spring and late fall; however, severe root damage and death of the plants may not develop until plant stress occurs from hot, dry weather. The development of take-all patch is more severe when the pH of the upper 1 inch of the soil is 6.5 and above.

Maintaining adequate levels of soil phosphorus and potassium has a suppressive effect on development of take-all patch. Types and rates of application of nitrogen-based fertilizers also impact on the severity of the disease. High rates of urea will enhance its development; however, recovery from the disease will be accelerated with applications of either ammonium sulfate or ammonium chloride. There appears to be no relationship between degree of thatch accumulation and the severity of take-all patch.

- **Control:**

1. Cultural Practices—Recovery of grass from take-all patch is usually slow; therefore, when small areas have been affected, the most practical immediate remedy is to remove the diseased turf and resod, or rake and reseed the affected areas.

Where the overall approach to control of take-all patch is concerned, management practices should be employed that establish and maintain acid soil conditions. Applications of sulfur and either ammonium sulfate or ammonium chloride in the early spring will reduce the severity of the disease. It is also important that a balanced fertilization program be followed. The potassium and phosphorous levels in the soil should not be allowed to become deficient.

2. Use of Fungicides—Azoxystrobin has been shown to be effective in the control of take-all patch. See Appendix Table I for a listing of the manufacturer and trade name for this fungicide.

Echinochloa crusgalli
(barnyard grass)

Rhizoctonia Yellow Patch

- **Pathogen:**
Rhizoctonia cerealis (Figure 3-1)
- **Grasses Affected:**
Annual bluegrass (*Poa annua*), Kentucky bluegrass (*Poa pratensis*), rough-stalk bluegrass (*Poa trivialis*), tall fescue (*Festuca arundinacea*), creeping bentgrass (*Agrostis palustris*), perennial ryegrass (*Lolium perenne*), bermudagrass (*Cynodon dactylon*), zoysiagrass (*Zoysia japonica*)
- **Season of Occurrence:**
Late winter, spring and fall
- **Symptoms and Signs:**
The symptoms of Rhizoctonia yellow patch can develop suddenly during cool, moist weather. Severity of the disease and types of symptom expression will vary somewhat depending on the turfgrass species involved and prevailing climatic conditions.

On bermudagrass and zoysiagrass, the development of the disease is usually limited to leaf yellowing. Annual bluegrass is highly susceptible to Rhizoctonia yellow patch. Under golf green management, annual bluegrass develops distinctive tan to straw-colored circles and/or patches 1 to 3 feet (0.3–0.9 m) in diameter (Plate 3-8). On cool season grasses under golf fairway, landscape or sports turf management, Rhizoctonia yellow patch is usually first seen as yellow, tan, or straw-colored patches ranging from 1 to 3 feet (0.3–0.9 m) in diameter. Individual patches may develop a pronounced sunken appearance that has been brought on by the decomposition of the thatch in the affected area. The grass in the center of the larger patches often recovers, leading to the formation of a frogeye pattern of patches of green plants with light yellow to tan outer borders (Plate 3-9).

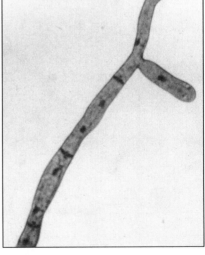

Figure 3-1. Hyphal strand of Rhizoctonia yellow patch pathogen. Note binucleate cells and right angle branching. *Courtesy Leon Burpee.*

Plate 3-8. Rhizoctonia yellow patch of annual bluegrass golf green.

Plate 3-9. Rhizoctonia yellow patch of Kentucky bluegrass.

In instances of field diagnosis on warm season grasses during cool, wet weather, it is important that one be able to distinguish between the symptom patterns of Rhizoctonia yellow patch and Rhizoctonia blight. With the warm season grasses, both diseases produce distinctively yellowed leaves and yellow patches of affected turf. However, patches of turf affected by the yellow patch pathogen will not have necrosis of lower leaf sheaths that is typical of Rhizoctonia blight.

• **Conditions Favoring Disease Development:**
The development of Rhizoctonia yellow patch is favored by cool, wet weather. The optimum air temperature range for disease development is 50–65 °F (10–18 °C). When the leaf symptoms are at the light yellow stage of development, the plants will recover if the air temperatures drop below 45 °F (7 °C) or go above 75 °F (24 °C). However, during times of extended rainfall, if the temperatures stay within the 50–65 °F (10–18 °C) range, foliar blighting will occur.

The earliest infections of the plants occur on the crowns and roots. However, the primary damage from the disease results from the direct infection and colonization of the foliage. The leaf yellowing symptom of the disease begins in the very early stages of tissue colonization by the pathogen.

• **Control:**
1. Cultural Practices—Good surface and subsurface drainage will reduce the severity of Rhizoctonia yellow patch by lowering the humidity within the canopy. Management practices that decrease the length of time the leaves are wet will aid in reducing the incidence of the disease. One such practice is the early morning removal of dew and guttation water from golf greens by poling or by dragging a water hose across them. The duration of the periods of daily leaf wetness can also be reduced by 2 to 4 hours by following a nighttime watering schedule in which the irrigation system is set to begin at least 3 hours after sunset and programmed to be completed before sunrise.

2. Use of Fungicides—Flutolanil is registered for the control of Rhizoctonia yellow patch. For a listing of the manufacturer and trade name of flutolanil, see Appendix Table I.

Rhizoctonia Blight
of Warm Season Turfgrasses

- **Pathogen:**

Rhizoctonia solani (Anastamosis group AG-2-2)

- **Grasses Affected:**

Bermudagrass (*Cynodon dactylon*), buffalograss (*Buchloe dactyloides*), centipedegrass (*Eremochloa ophiuroides*), St. Augustinegrass (*Stenotaphrum secundatum*), zoysiagrass (*Zoysia japonica*)

- **Season of Occurrence:**

Most destructive during spring and fall

- **Symptoms and Signs:**

On warm season grasses in the fall, Rhizoctonia blight is first seen in overall view as light green patches ranging from 2 inches (5 cm) to 2 feet (0.6 m) wide. Under conditions favorable for development of the disease, the color of these areas changes rapidly to a distinctive bright yellow and then to brown. In cases of prolonged outbreaks of Rhizoctonia blight, the diameter of diseased sections of turf may extend to 20 feet (6 m) or more. During extended periods of humid weather, these patches may develop dark purplish borders 2–6 inches (5–15 cm) wide. In the spring, large yellow patches become visible as the grasses resume growth from winter dormancy. These areas may rapidly turn to brown and continue to enlarge as long as relatively cool, wet weather persists (Plate 3-10; Plate 3-11; Plate 3-12).

Plate 3-10. Rhizoctonia blight of bermudagrass.

Plate 3-11. Rhizoctonia blight of zoysiagrass.

Plate 3-12. Rhizoctonia blight of St. Augustinegrass. *Courtesy Edward Freeman.*

A primary diagnostic feature of Rhizoctonia blight on zoysiagrass, bermudagrass, centipedegrass and St. Augustinegrass is a soft, dark brown to purplish rot of the lower portion of the leaf sheaths. This symptom pattern is most prevalent when the leaves are continuously wet for 48 hours or more. When periods of relatively dry weather follow major outbreaks of Rhizoctonia blight, a dry, reddish brown necrosis envelops the base of the leaf sheaths and spreads into the stem tissue (Plate 3-13; Plate 3-14).

In the advanced stages of disease development, an extensive soft rot will appear at the bases of the fascicles and the stems. When this happens, the dying plants at the edges of the patches can be easily pulled off the stolons.

• **Conditions Favoring Disease Development:**
Rhizoctonia blight of warm season grasses is most destructive during spring and fall months. The conditions that favor major outbreaks of the disease are extended periods of leaf wetness as the plants are going into dormancy in late fall or early winter, or when they are coming out of dormancy in the spring. The incidence of the disease is usually greater on turfgrass growing in low wet areas or on poorly drained soil. The sus-

Plate 3-13. Basal portion of zoysia leaf showing sheath rot and leaf yellowing symptoms of Rhizoctonia blight.

Plate 3-14. Sheath and culm necrosis symptoms of Rhizoctonia blight of St. Augustinegrass. *Courtesy Robert Haygood.*

ceptibility of plants to colonization by the Rhizoctonia blight pathogen is greater under conditions of high nitrogen fertilization.

• **Control:**

1. Cultural Practices—The maintenance of good drainage will reduce the incidence and severity of Rhizoctonia blight. Also, management practices that decrease the length of time the leaves are wet will aid in lowering the intensity of the disease. One procedure for reducing the leaf wetness periods is the early morning removal of water from bermudagrass golf greens by dragging a water hose across them or by poling. The practice of poling simply involves brushing the surface of the turf with long, limber bamboo poles.

The duration of the daily leaf wetness period can also be decreased 2 to 4 hours by following a nighttime watering schedule. Rather than irrigating immediately after sunrise or during late afternoon hours, the watering routine should be set to begin at least 3 hours after sunset and programmed to be completed before sunrise.

2. Use of Fungicides—The most effective fungicide programs for the control of Rhizoctonia blight of warm season grasses are those that begin the treatments with the first outbreak of the disease in the fall and continue the application schedule until the grass is into winter dormancy, and then initiate a fungicide application program in the same areas as the plants are coming out of dormancy in the spring.

Fungicides labeled for the control of Rhizoctonia blight of warm season turfgrasses include flutolanil, azoxystrobin, trifloxystrobin, chlorothalonil, iprodione, and quintozene. For a profile of these fungicides and a listing of representative trade names and manufacturers, see Appendix Table I.

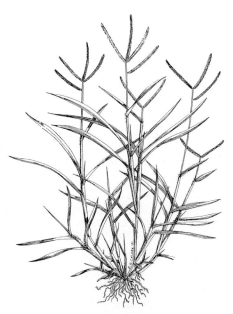

Eleusine indica
(goosegrass)

Corticium Red Thread

- **Pathogen:**

Laetisaria fuciformis (syn. *Corticium fuciforme*)

- **Grasses Affected:**

Annual bluegrass (*Poa annua*), creeping bentgrass (*Agrostis palustris*), bermudagrass (*Cynodon dactylon*), Kentucky bluegrass (*Poa pratensis*), hard fescue (*Festuca ovina* var. *duriuscula*), red fescue (*Festuca rubra*), sheep fescue (*Festuca ovina*), tall fescue (*Festuca arundinacea*), annual ryegrass (*Lolium multiflorum*), perennial ryegrass (*Lolium perenne*)

- **Season of Occurrence:**

Outbreaks of this disease commonly occur in late winter and early spring. However, during extended periods of rainfall and night temperatures in the high 60s to low 70s °F (19–22 °C), Corticium red thread can also cause severe damage to turfgrass during the summer months.

- **Symptoms and Signs:**

In overall view, Corticium red thread is seen as irregularly shaped patches of blighted turfgrass, ranging in size from 2 inches (5 cm) to 3 feet (0.9 m) in diameter. In cases of involvement of large areas of grass, the patches often have an overall "ragged" appearance due to a mixture of diseased and unaffected leaves (Plate 3-15). The disease is confined to the leaves and leaf sheaths only. Small, water-soaked spots develop

Plate 3-15. Corticium red thread of perennial ryegrass.

within 24–48 hours from the time of infection. Under favorable weather conditions, the affected leaves may be completely covered with the pink gelatinous growth of the pathogen.

A key feature for field diagnosis of Corticium red thread on tall cut grass is the presence of fine, threadlike, coral-pink structures (sclerotia) 1/16–1/4 inch (1–6 mm) in length at the terminal portions of the leaves (Plate 3-16).

Under the close mowing conditions of golf and bowling greens and golf tees, the affected areas range from 2–6 inches (5–15 cm) in diameter and are irregular in outline (Plate 3-17). Close examination of leaves from the affected areas will usually reveal the presence of a light reddish tinge to the sheaths. Also, although their numbers will be low because of the close mowing, the "red threads" that characterize the disease on tall cut grass can also be found in these areas.

• **Conditions Favoring Disease Development:**
The Corticium red thread pathogen is dispersed to new locations within a stand of turfgrass by diseased leaf tissue adhering to the surfaces of maintenance equipment. Although air temperatures in the 65–75 °F (18–24 °C) range are usually considered most conducive to the development of the disease, the length of time the surfaces of the leaves are wet appears to be an overriding factor in its development. Corticium red thread can develop on

Plate 3-16. Corticum red thread of perennial ryegrass, showing coral pink tendrils at ends of diseased leaves.

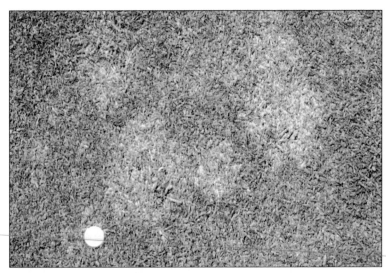

Plate 3-17. Corticum red thread creeping bentgrass under golf tee cutting height.

unfrozen turf under snow cover and during winter months that are marked by extended periods of rainfall. Also, major outbreaks of the disease can occur in warm summer months during prolonged rainfall.

Increased potassium fertilization has been shown to reduce the severity of the disease. Nitrogen nutrition does not lessen the susceptibility of turfgrass to Corticium red thread. However, application of nitrogenous fertilizers often reduces the severity of field symptoms of the disease. This is probably the result of more rapid plant recovery during periods of decreased activity on the part of the pathogen.

• **Control:**

1. Cultural Practices—If the stand of turfgrasses is under either low nitrogen or low potassium fertility, the application of muriate of potash and a readily available form of nitrogen fertilizer during periods of high disease incidence will aid in offsetting disease damage as well as facilitate faster plant recovery when the weather conditions are no longer favorable for the development of the disease.

2. Use of Fungicides—A preventive fungicidal application should be made when daytime air temperatures stabilize in the 60–70 °F (16–21 °C) range and rainfalls are frequent. Long term control of Corticium red thread can be accomplished by applications of either azoxystrobin, flutolanil or trifloxystrobin. The trade names and manufacturers of these fungicides are listed in Appendix Table I.

Limonomyces Pink Patch

- **Pathogen:**

Limonomyces roseipellis (Figure 3-2)

- **Grasses Affected:**

Bentgrasses (*Agrostis* spp.), perennial ryegrass (*Lolium perenne*), Kentucky bluegrass (*Poa pratensis*), annual bluegrass (*Poa annua*), creeping red fescue (*Festuca rubra*), and bermudagrass (*Cynodon dactylon*)

- **Season of Occurrence:**

Normally considered to be a disease of late winter and early spring. However, during periods of extended rainfall and night temperatures in the high 60s to low 70s °F (19–22 °C), Limonomyces pink patch can also cause severe damage to turfgrass during the summer months.

- **Symptoms and Signs:**

Limonomyces pink patch is confined to the above-ground plant parts. Symptoms are usually seen first along the margins of the leaves as small, irregularly shaped blotches of pink color bordered by light green to yellow bands of discolored leaf tissue. Eventually, the entire width of the leaf takes on a distinctive pinkish cast. When this occurs, a light brown to tan tip die back of the leaves develops. A thin, pink layer of fungal growth is often present on the surfaces of the leaves of affected plants. Finally, when conditions are particularly favorable for development of

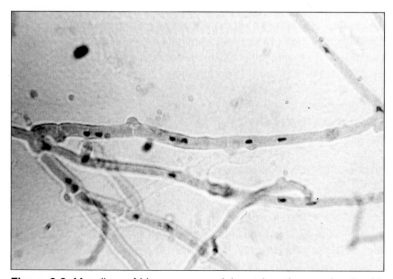

Figure 3-2. Mycelium of Limonomyces pink patch pathogen showing binucleate cells and clamp connections. *Courtesy Nicole O'Neil.*

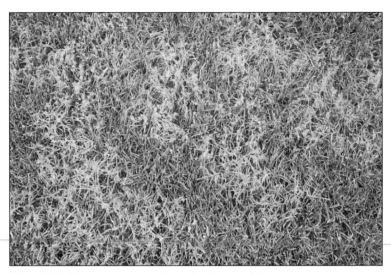

Plate 3-18. Limonomyces pink patch of perennial ryegrass.

the disease, numerous, distinctive pink-colored tufts of mycelium measuring 1/4 to 1/2 inch (0.6–1.3 cm) develop on the stems and leaf sheaths (Plate 3-18; Plate 3-19).

The patches often assume an overall pinkish tinge. When the grass is mowed frequently and growing under optimum nitrogen fertilization, the patches seldom reach more than 20 inches (0.6 m) in diameter and overall damage is minimal. However, on turf that is mowed infrequently and is under low nitrogen fertilization, damage from the disease may be severe. In these instances, all of the above-ground plant parts may become completely blighted.

Limonomyces pink patch has certain symptoms in common with Corticium red thread; however, the fine, threadlike coral-pink tendrils that develop on leaves affected with Corticium red thread are never present in cases of Limonomyces pink patch. If these reddish tendrils are not present, one can be reasonably certain that the disease is Limonomyces pink. However, the two diseases can develop simultaneously in the same stand of grass. If the reddish sclerotia have formed at the tips of the leaves and pinkish tufts of mycelium are also present, a laboratory-based comparison of the fungi colonizing the leaves will be necessary to determine if both diseases are present.

*   **Conditions Favoring Disease Development:**

The Limonomyces pink patch pathogen grows saprophytically on turfgrass debris. The fungus survives adverse climatic conditions by means

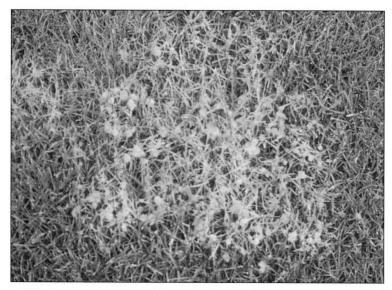

Plate 3-19. Limonomyces pink patch of perennial ryegrass. Note tufts of fungal growth.

of dormant mycelium in colonized turfgrass debris, and as small, reddish, waxy pads of mycelium on the surfaces of leaves and stems. Leaves in advanced stages of senescence are more vulnerable to infection and colonization than young, actively growing leaves. As the disease progresses, the older leaves often become bleached and matted together, while the younger leaves maintain their integrity.

The development of Limonomyces pink patch is favored by extended periods of leaf wetness and air temperatures in the 60–70 °F (16–21 °C) range. Also, the disease is more severe under low nitrogen fertilization. Typically, Limonomyces pink patch is considered to be a minor disease, however, on semidormant turf, during extended periods of rainfall the above ground plant parts can be killed back to the tillers. Also during prolonged periods of wet weather, Limonomyces pink patch can be very severe on infrequently mowed grass growing under low nitrogen fertilization.

• **Control:**
Long-term control of Limonomyces pink patch can be accomplished by applications of either azoxystrobin or flutolanil. A preventive fungicidal application should be made when daytime air temperatures stabilize in the 60–70 °F (16–21 °C) range and rainfalls are frequent. The trade names and manufacturers of these fungicides are listed in Appendix Table I.

SUMMER PATCH DISEASES
Fusarium Blight

- **Pathogens:**

Fusarium culmorum (syn. *F. roseum* f. sp. *cerealis*); *Fusarium poae* (syn. *F.tricinctum* f.sp. *poae*) (Figure 3-3)

- **Grasses Affected:**

Bentgrasses (*Agrostis* spp.), annual bluegrass (*Poa annua*), Kentucky bluegrass (*Poa pratensis*), centipedegrass (*Eremochloa ophiuroides*), Chewing's fescue (*Festuca rubra* var. *commutata*), red fescue (*Festuca rubra*), tall fescue (*Festuca arundinacea*), hard fescue (*Festuca ovina* var. *duriuscula),* annual ryegrass (*Lolium multiflorum*), perennial ryegrass (*Lolium perenne*)

- **Season of Occurrence:**

Late spring and summer

- **Symptoms and Signs:**

Leaf lesions first appear as irregularly-shaped, dark green blotches. They fade rapidly to a light green, then assume a reddish brown hue, and finally

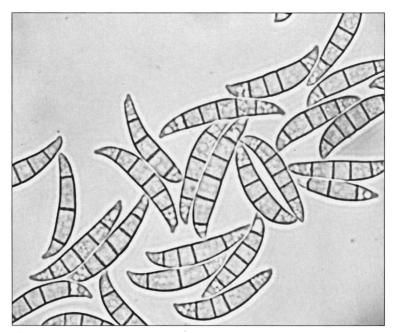

Figure 3-3. Conidiospores of Fusarium blight pathogen, *Fusarium culmorum. Courtesy Paul E. Nelson.*

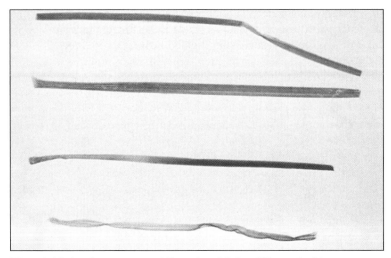

Plate 3-20. Leaf symptoms of Fusarium blight of Kentucky bluegrass.

become dull tan. Individual lesions often involve the entire width of the leaf blade and may extend up to 1/2 inch (6 mm) in length (Plate 3-20).

In overall view, the affected areas first show scattered light green patches 2–6 inches (5–15 cm) in diameter. Under environmental conditions favorable for disease development, the color of these patches changes in a 36–48 hour period to a dull reddish brown, then to tan, and finally to a light straw color. Initially, the shapes of the patches are elongate streaks, crescents, or circular (Plate 3-21).

Plate 3-21. Fusarium blight of Kentucky bluegrass.

The most characteristic feature of the total syndrome is seen in the later stages of disease development. At these times, there are present more or less circular patches of blighted turfgrass 1 to 3 feet (0.3–0.9 m) in diameter. Light tan to straw colored, they often have reddish brown margins 1 to 2 inches (3–5 cm) wide and contain center tufts of green, apparently unaffected grass (Plate 3-22; Plate 3-23; Plate 3-24). This combination produces a distinctive frogeye effect. When optimum conditions for disease development exist for an extended period of time, these affected areas coalesce. As a result, large areas of turfgrass may be blighted (Plate 3-25). During periods of relatively high soil moisture brought on by frequent rainfalls, the pinkish growth of the pathogen can be seen on the root and crown tissue near the soil surface (Plate 3-26).

•   **Conditions Favoring Disease Development:**
In landscape turf, the first part of the sward to show visible symptoms of the disease is usually sloping land that is close to a paved walk, driveway or parking lot. Turf on south-facing slopes will usually show symptoms of Fusarium blight earlier and become more severely diseased than sod on slopes with northern exposures. Also, there is a tendency for the disease to grade in severity in proportion to the intensity and duration of sunlight. Areas in full sunlight throughout the daytime

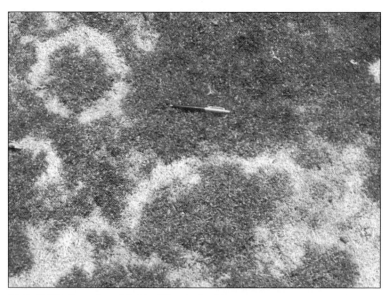

Plate 3-22. Fusarium blight of Kentucky bluegrass.

Plate 3-23. Fusarium blight of Kentucky bluegrass.

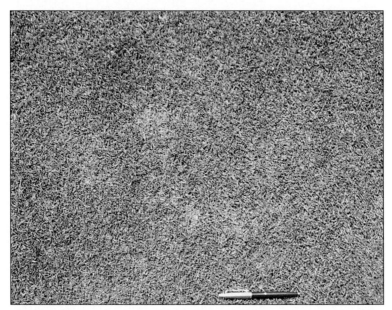

Plate 3-24. Fusarium blight of bentgrass under golf green management.

Plate 3-25. Fusarium blight of tall fescue. *Courtesy Philip Colbaugh.*

hours will usually be more severely affected by Fusarium blight than those in which there is shading from the sun during the noon to 4 p.m. period of the day.

Fusarium blight is more severe on turf under high nitrogen nutrition. High soil moisture stress also increases the severity of the disease. Heavy thatch [thickness greater than 1/2 inch (1.3 cm)] is conducive to heavy outbreaks of Fusarium blight. Optimum atmospheric conditions for the development of the disease are extended periods of dry weather combined with ground fogs or heavy dews which establish and maintain free water on the leaf surfaces until late morning hours.

• **Control:**

1. Cultural Practices—Thatch accumulation should not be allowed to exceed 1/2 inch (1.3 cm) thickness. Areas in which soil compaction is a problem should be subjected to core cultivation. When feasible, clippings should be removed. During periods of high air temperatures, irrigation should be employed frequently enough to maintain the soil at field capacity (-0.033 MPa). High applications of nitrogen in the spring should be avoided.

2. Use of Fungicides—Fusarium blight may be controlled by the use of either triadimefon, fenarimol, or thiophanate methyl. For a listing of rep-

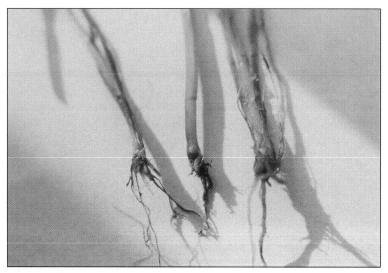

Plate 3-26. Symptoms of Fusarium blight of crown and root tissue of fall fescue. *Courtesy Philip Colbaugh.*

resentative trade names and manufacturers of these fungicides, see Appendix Table I.

In turf with a history of recurring Fusarium blight, the initial fungicide application should be made immediately after the first occurrence of a night temperature that does not drop below 70 °F (21 °C) and continued on a 10–12 day schedule as long as this condition persists.

Summer Patch

- **Pathogen:**

Magnaporthe poae

- **Grasses Affected:**

Annual bluegrass (*Poa annua*), creeping bentgrass (*Agrostis palustris*), Kentucky bluegrass (*Poa pratensis*), hard fescue (*Festuca ovina* var. *duriuscula*), red fescue (*Festuca rubra*), Chewing's fescue (*Festuca rubra* var. *commutata*), perennial ryegrass (*Lolium perenne*)

- **Season of Occurrence:**

Summer

- **Symptoms and Signs:**

There are no distinctive leaf lesions associated with summer patch. In overall view, the disease is first seen on fairways and landscape turf as gray-green, wilted patches 4–6 inches (10–15 cm) in diameter. These areas fade rapidly to a light brown color as the leaves wither and then die. The individual patches may expand up to 12 inches (30 cm). These areas often coalesce, involving large sections of turf. There may be locations within affected stands that exhibit the basic frogeye pattern of more or less circular patches of dead grass with center tufts of apparently unaffected plants (Plate 3-27).

On golf greens and bowling greens with high populations of annual

Plate 3-27. Summer patch of Kentucky bluegrass. *Courtesy Noel Jackson.*

bluegrass, summer patch is usually first seen as somewhat circular, reddish-brown patches 2–3 inches (5–8 cm) in diameter. These individual patches may progress to 10 to 12 inches (25–30 cm). Bentgrasses typically grow in the centers of the patches, creating a frogeye pattern. When conditions are particularly favorable for the development of the disease, the individual patches will coalesce and envelope large areas of turf (Plate 3-28).

The roots and crowns of diseased plants often contain blackened, necrotic tissue. In advanced stages of disease development, entire roots and crowns may be blackened. Also, dark brown "runner hyphae" are often present on the surface of the crowns and roots (Figure 3-4).

- **Conditions Favoring Disease Development:**

The development of summer patch is favored by hot humid weather. Active colonization of turfgrass roots and crowns by the pathogen begins at air temperatures in the 65–70 °F (18–21 °C) range; however, visible symptoms usually do not become evident until the temperatures reach 85 °F (30 °C). The severity of the disease is greatest at air temperatures between 85–95 °F (30–35 °C).

Summer patch is more severe on Kentucky bluegrass cut at 1 1/2 inches (4 cm) than at 2 1/4 inches (6 cm). Turfgrasses with restricted root systems are more prone to severe outbreaks of summer patch than turf

Plate 3-28. Summer patch of annual bluegrass under golf green management. *Courtesy Bruce Clark.*

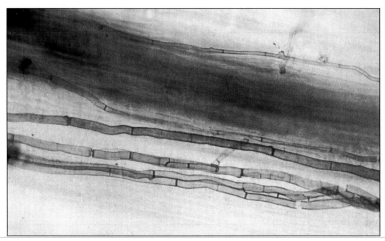

Figure 3-4. Runner hyphae of summer patch pathogen on annual blue-grass root. *Courtesy Bruce Clarke.*

with well developed roots. The disease is more severe on compacted soils and in areas characterized by recurrent foot traffic (such as the margins of golf and bowling greens and small golf greens) and under alkaline soil conditions. Also, the development of summer patch is usually greater at high soil moisture levels.

• **Control:**

1. Cultural Practices—Core aerification and improving drainage on compacted and poorly drained soils will lessen the intensity of the disease. Also, at the appearance of drought symptoms, irrigation schedules should be implemented that maintain adequate moisture to the depth of rooting. In annual bluegrass-dependent situations, during periods of heat stress, summer patch damage can be lessened to some extent by syringing and raising the cutting height.

2. Use of Fungicides—Azoxystrobin, trifloxystrobin, propiconozole, triadimefon, fenarimol, and thiophanate methyl have been reported to control summer patch. For a listing of representative trade names and manufacturers of these fungicides, see Appendix Table I.

In areas with a history of recurring summer patch, in the spring the soil temperature should be monitored at a depth of 2 to 4 inches (5 cm) during the hottest part of the day (2:00–3:00 p.m.). The first treatment should be made when this temperature reaches 65 °F (18 °C) on two consecutive days. This application should be followed with two additional treatments at 4-week intervals.

Pythium Blight

- **Pathogens:**

Pythium aphanidermatum, P. arrhenomanes, P. graminicola, P. myriotylum, P. ultimum (Plate 3-29)

- **Grasses Affected:**

Annual bluegrass (*Poa annua*), Kentucky bluegrass (*Poa pratensis*), roughstalk bluegrass (*Poa trivialis*), bermudagrass (*Cynodon dactylon*), bentgrasses (*Agrostis* spp.), tall fescue (*Festuca arundinacea*), red fescue (*Festuca rubra*), annual ryegrass (*Lolium multiflorum*), perennial ryegrass (*Lolium perenne*), St. Augustinegrass (*Stenotaphrum secundatum*)

- **Season of Occurrence:**

Late spring, summer, and early fall

- **Symptoms and Signs:**

In overall view, Pythium blight is first seen as small, irregularly shaped, purplish areas ranging from 1 to 4 inches (2.5–10 cm) in diameter. The individual leaves in these patches have a dark, water-soaked appearance As colonization by the fungus progresses, they become soft and slimy, and when they are in contact with each other, they mat together (Plate 3-30).

In the early morning hours, or if conditions of high humidity exist throughout the day, the leaves of diseased plants may be covered with the white, cobwebby, mycelium of the pathogen. Also during these times the older patches often develop dark purplish borders up to 1 inch (2.5 cm) wide (Plate 3-31; Plate 3-32; Plate 3-33).

The color of the affected leaves soon changes to light brown or reddish brown, and they become dry and shriveled. In the event the growth of the pathogen is checked before the entire leaf is colonized, distinct straw-colored lesions of varying size will develop. Generally, these lesions will not have distinct borders be-

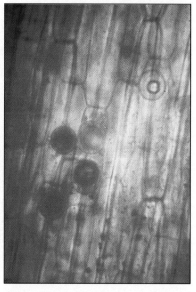

Plate 3-29. Oospores of Pythium blight pathogen embedded in bentgrass leaf.

Plate 3-30. Pythium blight of bentgrass golf green.

Plate 3-31. Mycelium of Pythium blight pathogen growing on creeping bentgrass leaves.

Plate 3-32. Pythium blight of creeping bentgrass under golf green management.

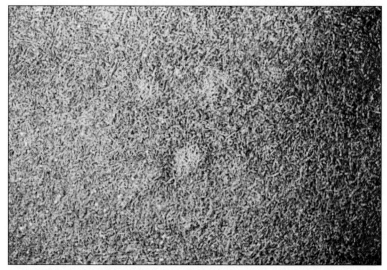

Plate 3-33. Pythium blight of creeping bentgrass under golf green management. Note purplish borders on the patches.

tween healthy and diseased tissue, in contrast to Sclerotinia dollar spot or Rhizoctonia blight.

Blighting of the foliage within the developing patches may be uniform, or the affected areas may develop as frogeyes—circles of blighted grass with centers of green, apparently healthy plants. Individual patches of affected grass frequently coalesce to envelop sections of turf ranging from 1 to 10 feet (0.3–3 m) in diameter (Plate 3-34; Plate 3-35). At times, the affected areas may develop as elongate streaks (Plate 3-36). Development of this pattern of blighting is apparently the result of the pathogen being washed over the surface of the soil or "tracked" by mowers.

• **Conditions Favoring Disease Development:**
The pathogen is spread to new locations by transport of diseased leaves on mowing and cultivation equipment and walkers' shoes, and fungal spores and diseased leaf fragments in flowing surface water.

The highest frequency of infections occur during extended periods of leaf wetness brought on by either rainfall, high atmospheric humidity in the leaf zone, dew formation, or night and morning ground fogs. Weather conditions that normally precede outbreaks of severe Pythium blight are (a) daytime air temperatures of 86 °F (30 °C) or greater, and (b) nighttime air temperatures of 68 °F (20 °C) or above in combination with 15 or more consecutive hours in which the relative humidity is 90 percent or higher.

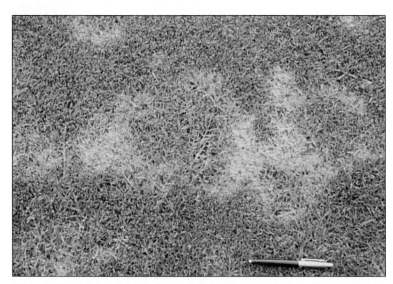

Plate 3-34. Pythium blight of creeping bentgrass under golf tee management. Note the presence of frogeye patches.

Plate 3-35. Pythium blight of creeping bentgrass under golf green management.

Plate 3-36. Pythium blight of creeping bentgrass under golf green management. This pattern is caused by the coalescence of several individual patches.

Bentgrass is more susceptible to Pythium blight when grown at low soil moisture contents, as compared with plants held near field capacity (-0.033 MPa) by frequent irrigations. Also, development of Pythium blight is greatest when the plants are growing under high nitrogen fertilization.

- **Control:**

1. Cultural Practices—Maintain satisfactory, but not luxuriant, plant growth through the use of balanced fertilizer applications. Also, decrease the length of time the leaves are wet by poling or dragging a water hose across the turf in the early morning. Duration of the periods of daily leaf wetness can also be reduced by 2 to 4 hours by programming a nighttime watering schedule in which the irrigation begins at least 3 hours after sunset and is completed before sunrise.

2. Use of Fungicides—Pythium blight may be controlled by applications of either mefenoxam, propamocarb, fosetyl Al, ethazole, chloroneb, trifloxystrobin, azoxystrobin, or mancozeb. The trade names and representative manufacturers of each of these fungicides are given in Appendix Table I. The synergistic fungicide combinations listed in Table 3–1 are particularly effective in the control of Pythium blight.

In stands of turfgrass with a history of Pythium blight, the first fungicide application should be made immediately after the first occurrence of night temperatures that do not drop below 65 °F (18 °C) and the relative humidity during the nighttime period under consideration is 85 percent or higher.

Table 3-1. Synergistic combinations of fungicides for increased effectiveness in the control of Pythium blight. *From Couch,1995.*

Fungicide Combinations[a, b]	Rate per 1000 ft^2 (93m^2)	
	Formulated Product	Active Ingredient
Mancozeb (80 WP)	4.0 oz (113.4 g)	3.2 oz (90.7 g)
+ Mefenoxam (21.3 EC)	0.5 oz (14.8 ml)	0.1 oz (3.0 ml)
Mancozeb (80 WP)	4.0 oz (113.4 g)	3.2 oz (90.7 g)
+ Propamocarb (6 F)	1.3 oz (38.5 ml)	1.0 oz (28.4 ml)
Fosetyl Al (80 WP)	4.0 oz (113.4 g)	3.2 oz (90.7 g)
+ Propamocarb (6 F)	1.3 oz (38.5 ml)	1.0 oz (28.4 ml)

[a]For a listing of the trade names for each of these fungicides, see Appendix Table I.
[b]The information given in parenthesis after each fungicide indicates formulation type and amount of active ingredient.

Rhizoctonia Blight
of Cool Season Turfgrasses

• **Pathogen:**

Rhizoctonia solani (Figure 3-5)

• **Grasses Affected:**

Annual bluegrass (*Poa annua*), bentgrasses (*Agrostis* spp.), Kentucky bluegrass (*Poa pratensis*), roughstalk bluegrass (*Poa trivialis*), Chewing's fescue (*Festuca rubra* var. *commutata*), red fescue (*Festuca rubra*), sheep fescue (*Festuca ovina*), tall fescue (*Festuca arundinacea*), annual ryegrass (*Lolium multiflorum*), perennial ryegrass (*Lolium perenne*)

• **Season of Occurrence:**

Late spring and summer

• **Symptoms and Signs:**

Symptom patterns of Rhizoctonia blight vary with the turfgrass type, height of cut and prevailing weather conditions. Under close mowing, as practiced for golf and bowling greens, the disease appears as irregularly shaped patches of blighted turfgrass ranging from 2 inches (5 cm) to 2 feet (0.6 m) or more in diameter. The initial color of these patches is usually a purplish green, but soon fades to light brown. During periods of

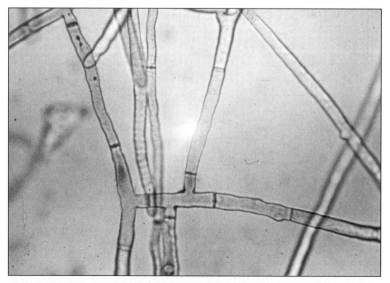

Figure 3-5. Mycelium of Rhizoctonia blight pathogen showing right angle branching of hyphal strands. *Courtesy Leon Burpee.*

warm, humid weather, dark, purplish "smoke rings" 1/2 to 2 inches (1.3–6 cm) may border the individual patches (Plate 3-37; Plate 3-38).

On landscape and sports turf and golf course fairways, Rhizoctonia blight is first seen as irregularly shaped areas of blighted grass ranging from a few inches to several feet in diameter. Initially, leaf death is usu-

Plate 3-37. Rhizoctonia blight of creeping bentgrass under golf green management.

Plate 3-38. Rhizoctonia blight of perennial ryegrass.

ally uniform throughout the affected areas. As the disease progresses, many of the affected areas may take on the form of circular patches of dull tan to brown grass 1 to 3 feet (0.3–0.9 m) in diameter, with center sections of green, unaffected plants. Eventually, the individual patches may coalesce to form irregularly shaped areas of uniformly blighted grass up to 50 feet (15 m) in diameter (Plate 3-39; Plate 3-40; Plate 3-41).

Plate 3-39. Rhizoctonia blight of tall fescue.

Plate 3-40. Rhizoctonia blight of tall fescue.

Plate 3-41. Rhizoctonia blight of Kentucky bluegrass.

Individual leaf symptoms of Rhizoctonia blight on Kentucky blue-grass, tall fescue and ryegrass first show as small, dull tan lesions. As these lesions enlarge, they develop reddish-brown margins. Eventually they may expand to envelop large sections of the leaf. At this stage of disease development, the entire leaf may become necrotic and take on a light brown appearance. The affected leaves become more-or-less brittle in texture, but they usually retain their original shape (Plate 3-42; Plate 3-43).

• **Conditions Favoring Disease Development:**
Outbreaks of Rhizoctonia blight of cool season turfgrasses are favored by high air temperatures in conjunction with extended periods of leaf wetness. The optimum conditions for a high frequency of infections are continuously wet leaves for 48 hours or more and air temperatures in the high 70s to low 80s F. Soil moisture levels in the readily available range [field capacity (-0.033 MPa) to permanent wilting percentage (-1.5 MPa)] do not influence the development of Rhizoctonia blight. However, the severity of the disease is much greater in turfgrass grown at high nitrogen with normal phosphorous and potassium levels than when the plants are grown at normal, balanced fertility.

• **Control:**
1. Cultural Practices—Short-term increases in the nitrogen fertilization of a bentgrass golf and bowling greens during weather conditions that are conducive to outbreaks of Rhizoctonia blight should be accom-

Plate 3-42. Leaf symptoms of Rhizoctonia blight of tall fescue.

panied by an increase in the dosage levels and frequency of application of fungicides.

Decreasing the length of time the leaves are wet will aid in reducing the incidence of Rhizoctonia blight. One such practice is the early morning removal of dew and guttation water from putting greens by poling or by dragging a water hose across them. The duration of the periods of daily leaf wetness can also be reduced by 2 to 4 hours by following a nighttime watering schedule in which the irrigation system is set to begin at least three hours after sunset and programmed to be completed before sunrise.

2. Use of Fungicides—Rhizoctonia blight of cool season turf-

Plate 3-43. Leaf symptoms of Rhizoctonia blight of Kentucky bluegrass.

grasses may be controlled with applications of flutolanil, trifloxystrobin azoxystrobin, iprodione, chlorothalonil, myclobutanil, or mancozeb. The trade names and representative manufacturers of each of these fungicides are given in Appendix Table I.

In stands of turfgrass with a history of Rhizoctonia blight, the first fungicide treatment should be made immediately after the first occurrence of night temperatures that do not drop below 70 °F (21 °C) and the relative humidity during the nighttime period under consideration is 85 percent or higher.

Lolium temulentum
(annual ryegrass, bearded darnel, cheat, darnel, ivray,
poison darnel, poison ryegrass, white darnel)

Sclerotinia Dollar Spot

- **Pathogen:**

Sclerotinia homoeocarpa

- **Grasses Affected:**

Bahiagrass (*Paspalum notatum*), bentgrasses (*Agrostis* spp.), bermudagrass (*Cynodon dactylon*), annual bluegrass (*Poa annua*), Kentucky bluegrass (*Poa pratensis*), centipedegrass (*Eremochloa ophiuroides*), sheep fescue (*Festuca ovina*), red fescue (*Festuca rubra*), tall fescue (*Festuca arundinacea*), zoysiagrass (*Zoysia japonica* and *Zoysia tenuifolia*)

- **Season of Occurrence:**

Late spring and summer

- **Symptoms and Signs:**

Individual leaves at first show yellow-green blotches, which progress to a water-soaked appearance, and finally bleach to a straw-colored tan with reddish brown borders. Entire leaves are commonly blighted, but, in some cases, only portions of leaves may become necrotic (Plate 3-44; Plate 3-45).

In overall view, the symptom pattern for Sclerotinia dollar spot varies with management practices. Under close mowing, as with golf

Plate 3-44. Leaf symptoms of Sclerotinia dollar spot of Kentucky bluegrass.

greens and bowling greens, the disease is first seen as very small spots of blighted turfgrass. These develop into circular, straw-colored patches 2 to 3 inches (5–7.5 cm) in diameter. The affected areas are usually sharply outlined against the surrounding, healthy turfgrass (Plate 3-46; Plate 3-47). In the early morning hours, while dew is still present on the leaves, a white

Plate 3-45. Leaf symptoms of Sclerotinia dollar spot of bermudagrass.

growth of mycelium may sometimes be seen covering the affected leaves. If progress of the disease is unchecked by fungicide applications, the individual patches frequently coalesce and involve large areas of turf.

Plate 3-46. Sclerotinia dollar spot of creeping bentgrass under golf green management.

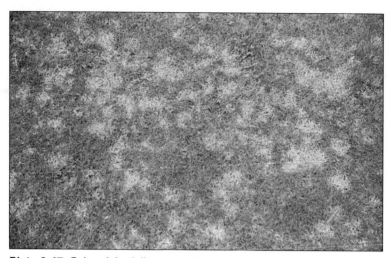

Plate 3-47. Sclerotinia dollar spot of bentgrass.

With the high mowing practices generally employed for landscape and sports turf and golf course fairways, the rather small, sharply outlined patches on putting and bowling greens are replaced by irregularly shaped, straw-colored areas of blighted turfgrass, ranging from 6 inches to 12 feet (15 cm-3.5 m) in diameter (Plate 3-48). In this form, the disease is some-

Plate 3-48. Sclerotinia dollar spot of Kentucky bluegrass.

times mistakenly diagnosed on landscape turf as drought injury, dull rotary mower injury, female dog injury, or fertilizer damage. During extended periods of high atmospheric humidity, the white mycelium of the pathogen will grow freely over the surface of the leaves and extend from leaf to leaf in a dense, cottony-type of formation (Plate 3-49).

• **Conditions Favoring Disease Development:**
The dollar spot pathogen overwinters in the form of resting bodies (sclerotia) and as dormant mycelium in the crowns and roots of diseased plants. When the temperatures in the area of the turf foliage reaches 60 °F (16 °C), the organism resumes growth. Disease development reaches its peak when the temperature in this zone ranges from 70–80 °F (21–27 °C) and the nighttime atmospheric humidity is 85 percent or higher.

Heavy thatch (thickness greater than 1/2 inch [1.3 cm]) is conducive to severe outbreaks of Sclerotinia dollar spot. Soil pH does not affect the severity of the disease. However, damage from Sclerotinia dollar spot is much greater on turfgrass growing under low soil moisture conditions. Also, as a general rule, damage from Sclerotinia dollar spot is more severe on turf grown under low nitrogen fertility.

• **Control:**
1. Cultural Practices—Removal of dew and guttation water by mowing or poling during early morning hours will significantly reduce in the

Plate 3-49. Mycelium of Sclerotinia dollar spot pathogen on leaves of perennial ryegrass.

incidence and severity of Sclerotinia dollar spot. During periods of high incidence of Sclerotinia dollar spot using an irrigation program that holds the soil at field capacity (-0.033 MPa) throughout the rooting zone will aid in the reduction of disease severity. Maintaining adequate levels of a balanced nutrition will also reduce incidence and severity of the disease and facilitate leaf recuperation.

2. Use of Fungicides—Sclerotinia dollar spot may be controlled by applications of either triadimefon, propiconazole, thiophanate-methyl, iprodione, vinclozolin, myclobutanil, fenarimol, or chlorothalonil. The trade names and representative manufacturers of each of these fungicides are given in Appendix Table I.

The synergistic fungicide combinations listed in Table 3–2 are particularly effective in the control of Sclerotinia dollar spot.

In a stand of turfgrass with a known history of Sclerotinia dollar spot occurrence, a preventive spray program at the lower label rate of the fungicide of choice should be initiated when the daytime air temperature stabilizes at approximately 70 °F (21 °C), and depending on the fungicide being used, continued at either 7–10 day or 14–21 day intervals. If the program is initiated after the appearance of symptoms, the first application should be made at the high label rate of the fungicide of choice. After which a low label rate application schedule may be employed.

Table 3-2. Fractional rate combinations of fungicides for the control of Sclerotinia dollar spot with synergy levels equivalent to that of the most effective component at its full label rate. *From Couch 1995.*

Fungicide Combinations[a, b]	Rate per 1000 ft² (93m²)	
	Formulated Product	Active Ingredient
Propiconozole (1.1 EC)	0.25 oz (7.4 ml)	0.035 oz (0.99 ml)
+ Triadimefon (25 WDG)	0.25 oz (7.1 g)	0.0625 oz (1.77 g)
Propiconozole (1.1 EC)	0.25 oz (7.4 ml)	0.035 oz (0.99ml)
+ Iprodione (2 F)	0.75 oz (22.2 ml)	0.1875 oz (5.32 ml)
Propiconozole (1.1 EC)	0.25 oz (7.4 ml)	0.035 oz (0.99 ml)
+ Chlorothalonil (4.17 FL)	1.5 oz (44.4 ml)	0.78 oz (22.17 ml)
Propiconozole (1.1 EC)	0.25 oz (7.4 ml)	0.035 oz (0.99 ml)
+ Vinclozolin (50 WP)	0.5 oz (14.2 g)	0.25 oz (7.09 g)

[a]For a listing of the trade names for each of these fungicides, see Appendix Table I.
[b]The information given in parenthesis after each fungicide indicates formulation type and amount of active ingredient.

Copper Spot

- **Pathogen:**
Gloeocercospora sorghi
- **Grasses Affected:**
Bentgrasses (*Agrostis* spp.), bermudagrass (*Cynodon dactylon*)
- **Season of Occurrence:**
Late spring and summer
- **Symptoms and Signs:**
Copper spot first appears as small light brown to reddish leaf lesions. As these spots enlarge, they become darker red. Under conditions favorable for disease development, the individual lesions often coalesce, thus blighting the entire leaf.

In the overall view, copper spot is seen as patches of salmon-pink, or copper-colored turfgrass, ranging from 1 to 3 inches (2.5–7.5 cm) in diameter. The intensity of coloration increases during wet weather due to the development of gelatinous, pink spore masses of the pathogen on the surface of the leaves (Plate 3-50; Plate 3-51).

Copper spot and Sclerotinia dollar spot sometimes occur simultaneously in the same stand of turfgrass. The two diseases can be distinguished from each other by differences in color and in sharpness of outline of the individual patches. Instead of a reddish-brown hue, the Sclerotinia dollar spot patches have a bleached, straw-colored appearance. Also, Sclerotinia

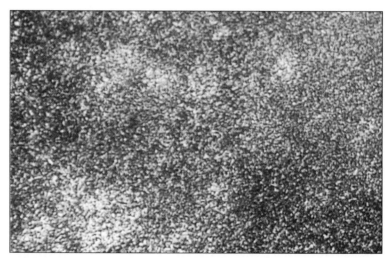

Plate 3-50. Copper spot of creeping bentgrass under golf green management.

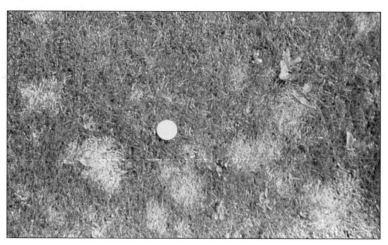

Plate 3-51. Copper spot of Colonial bentgrass under golf green management.

dollar spot patches usually have clearly defined borders, while the margins of copper spot patches tend to be diffuse (Plate 3-52).
• **Conditions Favoring Disease Development:**
The copper spot pathogen overwinters as sclerotia and thick-walled mycelium in the debris of the previous season's growth. Resumption of

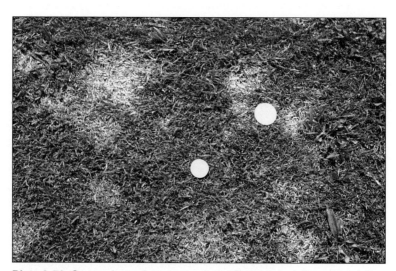

Plate 3-52. Comparison of copper spot and Sclerotinia dollar spot patches on creeping bentgrass under golf green management.

growth of the fungus occurs when the soil temperature at a depth of 1 inch (2.5 cm) remains at 63 °F (17 °C) or above for 7 days. Initial infections occur when the air temperatures reach 68 °F (20 °C). The severity of copper spot has been shown to be greater on bentgrass grown under high nitrogen fertilization.

Severe outbreaks of copper spot are favored by air temperatures in the 79–86 °F (26–30 °C) range in conjunction with extended periods of leaf wetness. The most favorable atmospheric moisture conditions for infection are those that facilitate the accumulation of free water on the leaves with a minimum of leaf washing, such as frequent, light showers, or the extended periods of heavy fog or the mists that are common to coastal areas. Under these air temperature-leaf wetness conditions, infection and colonization proceeds very rapidly. Leaf lesions are formed within 24 hours from the time of infection, and abundant sporulation takes place 24–48 hours later.

•   **Control:**
Copper spot may be controlled by applications of thiophanate-methyl, triadimefon, myclobutanil, chlorothalonil, vinclozolin, or iprodione. The trade names and representative manufacturers of each of these fungicides are given in Appendix Table I.

Preventive fungicidal applications should be made at 10–14 day intervals, beginning when the daytime air temperatures stabilize at 70–75 °F (21–24 °C). Applications at curative dosage levels should be made at 4–7 day intervals until total recovery has been achieved, after which 10 days may usually be allowed to elapse between applications.

Sclerotium Blight (Southern Blight)

- **Pathogen:**
Sclerotium rolfsii (Plate 3-53)
- **Grasses Affected:**
Annual bluegrass (*Poa annua*), bermudagrass (*Cynodon dactylon*), creeping bentgrass (*Agrostis palustris*), Kentucky bluegrass (*Poa pratensis*), annual ryegrass (*Lolium multiflorum*), perennial ryegrass (*Lolium perenne*)
- **Season of Occurrence:**
Spring and summer
- **Symptoms and Signs:**
Sclerotium blight is first seen on Kentucky bluegrass (*Poa pratensis*), as small, circular dead areas. Some green, apparently unaffected grass plants usually remain in the centers, thereby producing a frogeye appearance. These circular patches may enlarge up to 3 feet (0.9 m) in diameter. Some of the affected areas may develop into partial circles of arcs, rather than distinctive, circular patches (Plate 3-54). An unusual characteristic of the patches formed by Sclerotium blight is that weeds, such as clover, are also killed in the affected areas.

On bentgrass and/or annual bluegrass golf and bowling greens, Sclerotium blight first appears as yellowish crescent-shaped patches or circular rings with apparently healthy-looking grass in the center. The di-

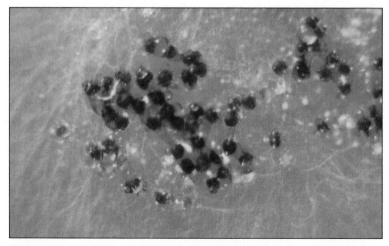

Plate 3-53. Sclerotia of the Sclerotium blight pathogen. *Courtesy Z. K. Punja.*

Plate 3-54. Sclerotium blight of Kentucky bluegrass. *Courtesy Leon Lucas.*

ameter of these areas will vary from 8 to 36 inches (0.2–0.9 m) (Plate 3-55). Although these patches may continue to enlarge at a somewhat steady rate throughout the growing season, the center portions of apparently healthy grass expands, but at a slower rate (Plate 3-56). Patches that are initiated during late summer usually show a uniform death of plants throughout. If turf does remain in the centers, it is usually chlorotic and sparsely populated.

During humid weather, masses of coarse white mycelium may grow on debris on the soil surface and on the dying grass at the edge of the patches. Also, small, white to brown, hard round bodies (sclerotia) 1/25 to 1/15 inch (1–2 mm) in diameter can frequently be seen on the dead grass or on the soil surface. The mycelium is not visible in dry weather, and sclerotia are difficult to find following periods of dry weather and later in the fall.

• **Conditions Favoring Disease Development:**
In the eastern United States, Sclerotium blight usually appears in mid-summer. In California, the disease becomes apparent in the late spring (usually the second or third week in May) and continues throughout the summer.

The Sclerotium blight pathogen survives adverse climatic conditions in the form of sclerotia in the debris of the previous season's growth

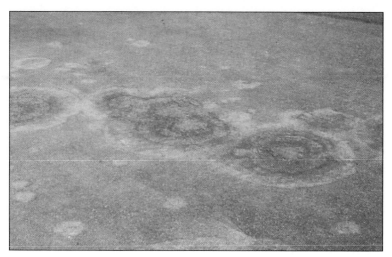

Plate 3-55. Sclerotium blight of creeping bentgrass. *Courtesy of Z. K. Punja & L. T. Lucas.*

or in the soil. Air temperatures above 75 °F (24 °C) and regularly occurring periods of thatch wetness followed by short drying periods of 1 to 2 hours are the conditions most conducive to sclerotial germination. Once the sclerotia have germinated, rapid and abundant growth of the mycelium is favored by continually moist thatch.

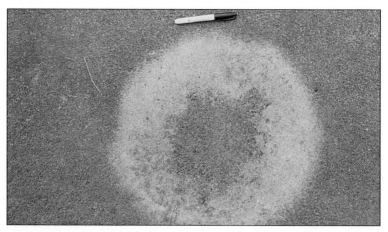

Plate 3-56. Sclerotium blight of creeping bentgrass under golf green management. *Courtesy Z. K. Punja.*

- **Control:**

Control of Sclerotium blight can be accomplished with applications of either triadimefon or flutolanil. The trade names and representative manufacturers of each of these fungicides are given in Appendix Table I.

In the eastern United states, treatments should begin in late June. In California, the first fungicide application should usually be made during the second or third week of May. Succeeding applications should continue at 2-week intervals for the remainder of the spring and summer months.

Panicum maximum
(guineagrass)

Melanotus White Patch

- **The Pathogen:**

Melanotus phillipsii

- **Grasses Affected:**

Creeping bentgrass (*Agrostis palustris*), tall fescue (*Festuca arundinacea*), red fescue (*Festuca rubra*), Chewing's fescue (*Festuca rubra* var. *commutata*)

- **Season of Occurrence:**

Summer

- **Symptoms and Signs:**

On fescue, Melanotus white patch is first seen as circular, white to off-white patches of blighted turfgrass ranging from 3 to 14 inches (8–36 cm) in diameter. Individual patches may be surrounded by a salmon-pink border. Also, the grass leaves within the affected areas may mat together and eventually become closely pressed to the soil surface. Under conditions favorable for disease development, the patches may coalesce to involve large areas of turfgrass (Plate 3-57; Plate 3-58).

Individual turfgrass blades are bleached to a light tan, starting at the tip and progressing toward the leaf sheath. The disease is restricted to the leaf blades, with the crowns of the plants remaining unaffected. Both the

Plate 3-57. Melanotus white patch of tall fescue.

Plate 3-58. Individual patch of Melanotus white patch of tall fescue.

mycelium and the fruiting bodies of the causal fungus occur commonly on the surfaces of the affected leaves. The mycelium develops as a grayish-white cobwebby growth on the leaves. The fruiting structures are very distinctive, and, therefore, serve as a valuable aid to diagnosis. These are small, grayish-white mushroomlike bodies, 1/16 to 5/16 inch (2–8 mm) in diameter. They develop initially as small round balls. Eventually, they open and the characteristic gills found on mushrooms are apparent on their lower sides (Plate 3-59).

On bentgrass golf and bowling greens Melanotus white patch is seen as irregularly shaped off-white patches of blighted turfgrass ranging from 2 to 6 inches (5–16 cm).

- **Conditions Favoring Disease Development:**

Melanotus white patch is more severe on tall fescue stand during their first year of establishment. Also, heavily seeded stands of grass are usually more severely affected by the disease than grass seeded at an appropriate rate.

Development of the disease is favored by hot, humid weather. Damage is greatest when the daytime air temperatures exceed 85 °F (30 °C) and the night temperatures do not fall below 70 °F (21 °C). In a given stand of grass, the most severe level of disease development occurs in the areas exposed to full sunlight rather than shaded areas. Also, outbreaks

Plate 3-59. Fruiting structures of Melanotus white patch pathogen on tall fescue leaves. *Courtesy Bobby Joyner.*

of Melanotus white patch are heaviest under conditions of low soil moisture content and low fertility.

• **Control:**

Tests with currently available turfgrass fungicides have failed to control this disease. Certain cultural practices, however, are helpful in reducing disease severity. Since Melanotus white patch is favored by hot, dry weather conditions, watering at frequent enough intervals to maintain the soil close to field capacity (-0.033 MPa) during these times will alleviate the problem to some extent. Also, although this information is usually of little solace to the owner of the fescue turf that is severely affected by this disease, since development of Melanotus white patch is dependent on high air temperatures and the causal fungus does not colonize crown tissue, with the return of cooler growing temperatures, total recovery of the affected plants often occurs.

WINTER PATCH DISEASES
Fusarium Patch (Pink Snow Mold)

- **The Pathogen:**

Microdochium nivale (syn. *Fusarium nivale*) (Figure 3-6)

- **Grasses Affected:**

Bentgrasses (*Agrostis* spp.), bermudagrass (*Cynodon dactylon*), annual bluegrass (*Poa annua*), Kentucky bluegrass (*Poa pratensis*), annual ryegrass (*Lolium multiflorum*), perennial ryegrass (*Lolium perenne*), Chewing's fescue (*Festuca rubra* var. *commutata*), sheep fescue (*Festuca ovina*), tall fescue (*Festuca arundinacea*), red fescue (*Festuca rubra*)

- **Season of Occurrence:**

Late fall, winter, and early spring

- **Symptoms and Signs:**

When outbreaks occur during cold, wet weather, Fusarium patch is first seen as roughly circular, water-soaked appearing spots 2 to 3 inches (5–7.5 cm) in diameter. The size of the patches may expand to 12 inches

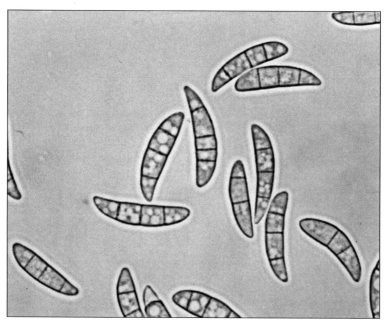

Figure 3-6. Conidiospores of Fusarium patch pathogen. *Courtesy Paul E. Nelson.*

(30 cm) or more in diameter within 48–72 hours from the onset of symptoms. Sometimes the plants in the center of a patch will begin recovering from the disease while the pathogen is still active at the periphery, producing a frogeye pattern. The color of the center portions of the developing patches changes to reddish-brown and then to tan. Adjacent patches often coalesce, resulting in a uniform blighting of large areas of turf (Plate 3-60).

The leaves of the affected plants tend to mat together. Also, a faint growth of white or dull pink mycelium may develop at the margins of the patches. On turf with a high population of annual bluegrass, the patches often develop distinctive reddish-brown borders 1/2 to 2 inches (1.3–5 cm) wide (Plate 3-61).

When Fusarium patch develops under a snow cover, as the snow melts, the prevailing overall symptoms are roughly circular, dull white patches from 3 to 12 inches (7.5–30 cm) in diameter.

• **Conditions Favoring Disease Development:**
In regions where snow cover is common, a deep, persistent snow that has been deposited on unfrozen ground establishes ideal conditions for development of Fusarium patch. However, snow cover is not a requirement for the development of the disease. In certain regions of the world, outbreaks of Fusarium patch can occur during any month of the year.

Plate 3-60. Fusarium patch of creeping bentgrass. *Courtesy Bobby Joyner.*

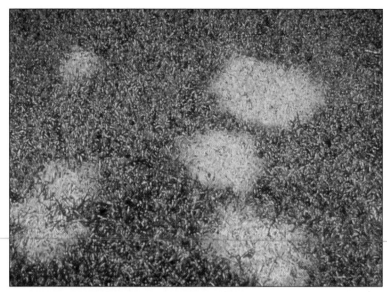

Plate 3-61. Fusarium patch of Kentucky bluegrass.

The optimum conditions for development of the disease are high atmospheric humidity and air temperatures ranging from 32–45 °F (0–7 °C). Under the conditions of extended leaf wetness brought on by ground fogs, mists, or frequent, light rain showers, severe outbreaks of Fusarium patch can also occur at air temperatures up to 65 °F (18 °C). When the air temperatures reach 70 °F (21 °C), the fungus ceases to be pathogenic. The severity of Fusarium patch is much greater when the pH of the top 1 inch (2.5 cm) of soil is 6.5 and above.

• **Control:**

1. Cultural Practices—In regions characterized by alkaline soils, every effort should be made to maintain the pH of the soil in the acid range. Where acidic soils exist, applications of lime should be made in the spring rather than late in the growing season.

The grass must not be left uncut in the fall. Raise the cutting height approximately 20 percent at the end of the growing season to allow for better cold temperature survival, and continue to mow until top growth is stopped.

2. Use of Fungicides—Fusarium patch may be controlled by single component applications of either triadimefon, propiconazole, thiophanate-methyl, quintozene, or fenarimol—or by the use of two or three component mixtures of quintozene, iprodione, chlorothalonil, or vinclozolin. The trade names and representative manufacturers of each of these fungicides are given in Appendix Table I.

In areas where winter snows are common, the treatment should be made within 2 weeks of the first predicted snowfall of the season, with midwinter and early spring applications when weather conditions permit. In regions with snow-free climates and the fungicide of choice is either triadimefon, propiconazole, thiophanate methyl, or fenarimol, treatments should begin just prior to the advent of cold, wet weather and continued at 7–14 day intervals as long as these weather conditions prevail.

Panicum repens
(torpedo grass, panic rampant)

Typhula Blight (Gray Snow Mold)

- **Pathogens:**

Typhula incarnata, Typhula ishikariensis

- **Grasses Affected:**

Bentgrasses (*Agrostis* spp.), tall fescue (*Festuca arundinacea*), red fescue (*Festuca rubra*), Chewing's fescue (*Festuca rubra* var. *commutata*), sheep fescue (*Festuca ovina*), perennial ryegrass (*Lolium perenne*), annual bluegrass (*Poa annua*), Kentucky bluegrass (*Poa pratensis*)

- **Season of Occurrence:**

Winter

- **Symptoms and Signs:**

Typhula blight is first seen in overall view as light yellow discolored turf 1 to 2 inches (2.5–5 cm) in diameter. The color of the leaves of the plants in these areas soon progress to a grayish white (Plate 3-62). In the latter stages of decomposition they become matted together. As these areas enlarge, a characteristic halo of grayish-white mycelial growth up to 1 inch (2.5 cm) in diameter develops at their advancing margin. Affected areas may measure up to 1 to 2 feet (30–60 cm) in diameter, but under optimum conditions for disease development, due to a merging of individual patches, zones of blighted turfgrass may be much larger (Plate 3-63).

Plate 3-62. Myelium of Typhula blight pathogen on creeping bentgrass leaves. Note area of charcoal-colored mycelium in upper left section of photo.

Plate 3-63. Typhula blight of creeping bentgrass.

A primary diagnostic feature of Typhula blight is the presence of characteristic sclerotia embedded in the leaves and crowns of diseased plants. These hardened fungus bodies range in size from an ordinary pin head to 3/16 inch (5 mm) in diameter. Early in the season, they are yellow or light brown. Eventually, they turn dark brown, are oval to spherical, and have a rough, irregular surface (Plate 3-64).

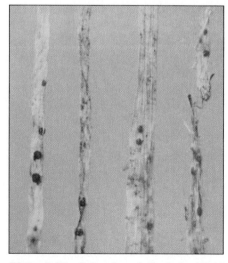

Plate 3-64. Sclerotia of Typhula blight pathogen embedded in creeping bentgrass leaves.

• **Conditions Favoring Disease Development:** The optimum conditions for development of Typhula blight are prolonged periods of high atmospheric humidity and air temperatures ranging from 36–40 °F (2–5 °C). Although mild cases of the disease are known to occur in regions where there is usually little or no snow cover, severe outbreaks normally develop only in areas in which the winter weather is characterized by persistent snow covers.

The Typhula blight pathogens survive the warm summer months in the form of sclerotia. In late fall, under the stimulus of cold weather, high humidity, and exposure to light rays of short wave length (2700 A to 3200 A), the sclerotia produce basidiospores. During cold, wet weather or under a snow cover, sclerotia germinate to produce mycelia. Light is not necessary for mycelial production. Although the spores are capable of producing infections, the vast majority of first infections are initiated by mycelium produced by the direct germination of sclerotia.

•   **Control:**

1. Cultural Practices—Late summer or early fall applications of nitrogenous fertilizers should be avoided. However, in the event of low soil fertility, in order to facilitate rapid plant regrowth in the spring moderate applications of balanced fertilizer may be made in late fall when the plants are entering into dormancy. Also, the installation of snow fences or the establishment of shrub or tree wind breaks to minimize snow accumulation in areas highly prone to drifts will help reduce the severity of Typhula blight.

Management practices that hold thatch at 1/2 inch (1.3 cm) or less will subsequently aid in reducing the severity of Typhula blight. Also, breaking up the matted turfgrass in the affected areas by brushing or raking before applying fungicides will facilitate faster plant recovery.

2. Use of Fungicides—Typhula blight may be controlled by single component applications of either flutolanil, propiconazole, or triadimefon—or the use of two or three component combinations of quintozene, flutolanil, vinclozolin, or iprodione, or a commercial mixture of oxycarboxin and thiram (Arrest 75 W™). The trade names and representative manufacturers of each of these fungicides are given in Appendix Table I.

The first fungicide application should be made within 2 weeks of the first predicted snowfall of the season. When weather conditions permit, fungicide applications should also be made at midwinter and early spring.

Sclerotinia Patch

- **Pathogen:**

Myriosclerotinia borealis (syn. *Sclerotinia borealis*)

- **Grasses Affected:**

Creeping bentgrasses (*Agrostis* spp.), red fescue (*Festuca rubra*), tall fescue (*Festuca arundinacea*), Kentucky bluegrass (*Poa pratensis*), perennial ryegrass (*Lolium perenne*)

- **Season of Occurrence:**

Winter

- **Symptoms and Signs:**

In overall view, the disease is first seen as small yellowish-green areas 2 to 4 inches (5–10 cm) in diameter. The leaves of the affected plants soon turn grayish white and become matted together. As these areas enlarge, a dark gray mycelial growth may be seen near the advancing margins. Individual patches may measure up to 24 to 36 inches (60–90 cm) in diameter. Under optimum conditions these often coalesce to create large areas of uniformly blighted turfgrass. Under severe disease conditions, crown and crown bud tissues become extensively rotted (Plate 3-65; Plate 3-67).

A key diagnostic feature of Sclerotinia patch is the presence of small, dull black sclerotia embedded in and on the surface of diseased

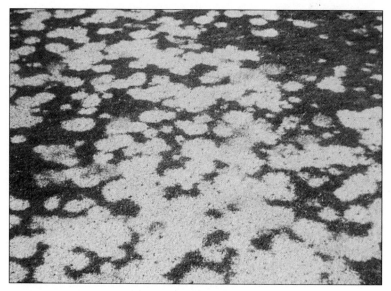

Plate 3-65. Sclerotinia patch of creeping bentgrass.

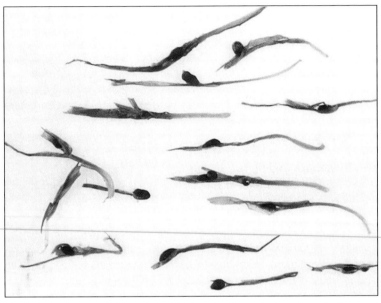

Plate 3-66. Sclerotia of Sclerotinia patch pathogen on surfaces of creeping bentgrass leaves. *Courtesy J. D. Smith.*

leaves. These are in contrast to the light yellow to reddish-brown sclerotia characteristic of Typhula blight (Plate 3-66).

• **Conditions Favoring Disease Development:**

The optimum weather conditions for the development of Sclerotinia patch are extended periods of frequent rainfalls and daily air temperatures ranging between 45–59 °F (6–15 °C), culminated by a deep, persistent snow cover that has been deposited on unfrozen ground. If the fall weather is dry and the air temperatures stabilize rapidly in the low 40s F, and/or the winter months are marked by light, infrequent snowfalls, the incidence and severity of the disease will be minimal.

The optimum temperature range for patch formation is 32–35 °F (0–2 °C). The reason for the heightened severity of Sclerotinia patch under deep, persistent snow cover is that the humid atmosphere and microclimate temperatures are ideal for the growth of the pathogen. Also, under these conditions, respiration continues, but the plants do not have sufficient light for photosynthesis. As the result, the carbohydrate reserves in the leaves become depleted, which in turn increases their susceptibility to infection and colonization by the fungus.

Applications of fertilizer high in water soluble nitrogen late in the

Plate 3-67. Sclerotinia patch of creeping bentgrass. Courtesy *J. D. Smith.*

growing season or immediately after snow melt in the spring will increase the severity of Sclerotinia patch. Development of the disease is also greater when the grass is grown at low levels of phosphorus or on highly acid soils.

• **Control:**

1. Cultural Practices—High rates of nitrogenous fertilizers in the late summer and fall should be avoided and soil phosphorous maintained at an adequate level. The soil pH should be maintained in the pH 5.6–6.0 range.

2. Use of Fungicides—Sclerotinia patch may be controlled with two applications of quintozene, benomyl, chlorothalonil, a mixture of oxycarboxin and thiram (Arrest 75 W^{TM}), or a mixture of quintozene and thiophanate methyl. The trade names and representative manufacturers of each of these fungicides are given in Appendix Table I.

The first fungicide application should be made in September and the second in October.

Cottony Snow Mold

- **Pathogen:**

Coprinus psychromorbidus

- **Grasses Affected:**

Bentgrasses (*Agrostis* spp.), annual bluegrass (*Poa annua*), Kentucky bluegrass (*Poa pratensis*), tall fescue (*Festuca arundinacea*), red fescue (*Festuca rubra*), Chewing's fescue (*Festuca rubra* var. *commutata*), sheep fescue (*Festuca ovina*)

- **Season of Occurrence:**

Winter

- **Symptoms and Signs:**

The overall symptom pattern for cottony snow mold is similar to that for Fusarium patch. Appearing with the resumption of plant growth at the first spring thaw, are irregularly shaped areas of grass that are pale yellow at first and then look bleached. Ranging generally up to 1 foot (0.3 m) in diameter, these areas may coalesce and involve large sections of turfgrass. Death of the entire plant usually results (Plate 3-68; Plate 3-69).

Cottony snow mold is distinguished from Fusarium patch by the presence of a mat of light gray mycelial growth over the affected areas instead of the pinkish growth characteristic of Fusarium patch (Plate

Plate 3-68. Cottony snow mold of Kentucky bluegrass. *Courtesy J. D. Smith.*

Plate 3-69. Cottony snow mold of creeping bentgrass. *Courtesy J. D. Smith.*

3-70). Also, the light brown sclerotia embedded in the leaves, the chief diagnostic feature of Typhula blight, are not present when the turf is affected by cottony snow mold.

- **Conditions Favoring Disease Development:**

Cottony snow mold is a disease of dormant plants. After winter dormancy has been established, a prolonged period of association, 45 to 60 days, is required before the plants are sufficiently predisposed to invasion by the pathogen. The incidence and severity of the disease are greater in turf that has received late season applications of high rates of nitrogenous fertilizers. Deep, persistent snow covers are not a requisite for the development of cottony snow mold. The optimum conditions for disease development are soil surface temperatures less than 32 °F (0 °C). When plant growth begins in the spring, the activity of the cottony snow mold pathogen ceases.

- **Control:**

1. Cultural Practices—If it is deemed necessary to apply an inorganic nitrogenous fertilizer after the end of July, the rate should not exceed 4 ounces actual nitrogen 1,000 square feet (0.1 kg N/92 m^2). However, a slow release organic nitrogen fertilizer can be put down with the fall fungicide application. This practice will bring about a good spring color and provide for the control of cottony blight.

Plate 3-70. Cottony snow mold of creeping bentgrass.

The use of polyethylene sheeting to cover the turf during the winter months has been shown to increase the effectiveness of fungicides in the control of cottony snow mold. Also, the installation of snow fences to minimize snow accumulation in areas highly prone to drifts will help reduce the severity of the disease.

2. Use of Fungicides—Cottony snow mold may be controlled with applications of quintozene, benomyl, chloroneb, or a commercial mixture of oxycarboxin and thiram (Arrest 75 W™). The trade names and representative manufacturers of each of these fungicides are given in Appendix Table I.

A fungicide treatment should be made soon after the first killing frost and a second application just prior to the first snowfall.

Spring Dead Spot of Bermudagrass

- **Pathogens:**

Depending on geographic location, one of four species of fungi functions as the incitant of this disease.

Australia: *Leptosphaeria narmari*

Unites States:

    Southeastern states—*Gaeumannomyces graminis* var. *graminis*

    California and Maryland—*Leptosphaeria korrae*

    Kansas—*Ophiosphaerella herpotricha*

- **Grasses Affected:**

Bermudagrass (*Cynodon dactylon*), African bermudagrass (*Cynodon transvaalensis*)

- **Season of Occurrence:**

Late fall, winter and early spring

- **Symptoms and Signs:**

The major symptoms of spring dead spot are first apparent when bermudagrass breaks winter dormancy and begins regrowth in the spring. In pure stands of bermudagrass, the disease is seen in overall view as very distinctive, depressed, well defined circular patches of straw-colored turf ranging from 6 inches (15 cm) to 3 feet (0.9 m) or more in diameter. In some instances, small patches will have coalesced and formed arcs (Plate 3-71; Plate 3-72; Plate 3-73).

Plate 3-71. Spring dead spot of bermudagrass.

Plate 3-72. Spring dead spot of bermudagrass under landscape management.

Small elliptically shaped black to dark brown lesions form on culm bases, crown buds, roots and stolons during the early stages of disease development. At the time of patch formation, the culm bases and crown buds develop a black to brown dry rot. The roots and stolons also become blackened and rotted and may be easily pulled loose from the parent plants. Strands of dark brown to black runner hyphae are often found on the surfaces of diseased roots and stolons.

If summer recovery does occur, outbreaks of spring dead spot often recur in many of the former patches during each successive winter. The diameter of these sites will continue to enlarge for several years. In these instances, the plants in the center of the patches sometimes survive, giving them a distinctive frogeye appearance—circles of blighted grass with centers of green, apparently healthy plants.

• **Conditions Favoring Disease Development:**
The fungi that incite spring dead spot infect and colonize the roots and stolons of bermudagrass in late summer or early fall when the prevailing daytime air temperatures are in the low to mid 70s F (21–24 °C). With the advent of colder fall and winter weather, the capacity of the bermudagrass to resist infections and compensate for the pathogenic effects of the fungus by producing new roots decreases. When the plants reach full dor-

Plate 3-73. Spring dead spot of bermudagrass under golf fairway management.

mancy, and the range of the daily air temperatures is 50 to 60 °F (10–15 °C) or less, extensive colonization of roots, crowns, and stolons by the fungus occurs.

Heavy thatch [thickness greater than 1/2 to 3/4 inch (1.3–2.0 cm)] is conducive to the development of spring dead spot. The severity of the disease is also greater in turf maintained at low cutting heights. Spring dead spot is more severe on bermudagrass grown at low potassium. Also, the use of nitrogenous fertilizers at high rates or the application of nitrogen late in the growing season may bring about an increase in the severity of the disease.

• **Control:**

1. Cultural Practices—The following management practices will reduce the severity of spring dead spot: The fertilization program should (a) use minimum rates of nitrogen for spring and summer growth, (b) avoid late growing season applications of nitrogen, and (c) maintain adequate soil potassium levels. Thatch should be kept between 1/2 and 3/4 inch (1.3–2.0 cm) in thickness, the turf mowed at the maximum height the use requirements will permit, and coring performed in the diseased areas to improve soil aeration and foster the production of new roots.

2. Use of Fungicides—Myclobutanil, propiconazole, azoxystrobin, and fenarimol and are registered in the United States for control of spring dead spot. The trade names and representative manufacturers of each of these fungicides are given in Appendix Table I.

Treatments may be made from September through November, however, the earlier application usually gives better results. The rates of application of fenarimol vary according to the month the treatment is made. The manufacturer's schedule should be consulted for the specific rate at the date of application in question.

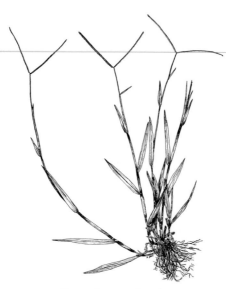

Paspalum conjugatum
(sour paspalum)

4

Leaf Spots

Introduction

All species of turfgrass are afflicted by leaf spot diseases. In the temperate climates of the world, during the spring, summer and fall seasons there is at least one leaf spot disease for each turfgrass species. In zones characterized by semi-arid climates, leaf spot diseases often develop on turfgrasses during the rainy season.

Development of this group of diseases is favored by extended periods of leaf surface wetness. Although outbreaks are commonly brought on by prolonged of rainfall, these diseases can also cause severe damage in situations of low precipitation but in which the leaf surfaces have been kept moist by either high day-night atmospheric humidity or night ground fogs that extend into the late morning hours. During continuous periods of leaf wetness, the blades and sheaths can become colonized so extensively by the invading fungus that the entire leaf dies. Also, in conjunction with the leaf spot feature of each disease a severe crown and root rot phase frequently develops which significantly reduces the vigor and drought tolerance of the plants.

Included in this section of the handbook are the major leaf spot diseases of turfgrasses. Each disease unit lists the name of the pathogen and the turfgrasses it parasitizes, describes the symptoms of the disease, gives the weather and management conditions most favorable for disease development, and outlines specific control procedures.

Melting-out of Kentucky Bluegrass

* **Pathogen:**

Drechslera poae (Figure 4-1)

* **Grasses Affected:**

Kentucky bluegrass (*Poa pratensis*), annual bluegrass (*Poa annua*), Canada bluegrass (*Poa compressa*), rough bluegrass (*Poa trivialis*)

* **Season of Occurrence:**

Spring and early summer

* **Symptoms and Signs:**

The disease occurs on all plant parts. On the leaves, it is first seen as minute, water-soaked lesions. These soon enlarge into dark, purplish-red, ovular areas 1/4 to 3/8 inch (6–9 mm) long and 1/16 to 1/8 inch (2–3 mm) wide. As the lesions enlarge, the color of the centers changes to brown, and finally to a dull white (Plate 4-1).

Lesions on the leaf sheath are generally not as regular in outline as those on the leaf blades, and the lighter colored center is usually missing. Colonization of the sheath tissue is often so extensive that the leaf is girdled at this point and drops from the plant. It is this leaf-dropping phase of the disease that has given rise to the name "melting-out." During severe outbreaks of the disease, bluegrass stands are commonly found with less than six leaves per square foot (0.3 m²) of turfgrass area.

Infection and colonization of the crowns, roots and rhizomes occurr in conjunction with the leaf lesioning phase of melting-out. The disease in these plant parts develops as a rot, appearing at first as a reddish-brown decay and finally turning dark brown to black as bacteria and other fungi begin to colonize the tissues.

* **Conditions Favoring Disease Development:**

Melting-out is a cool, wet weather disease. Optimum weather conditions for development of its leaf lesion phase are air temperatures in the 60–75 °F (16–24°C) range and extended

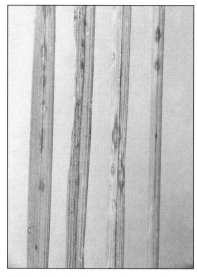

Plate 4-1. Leaf spot phase of melting-out of Kentucky bluegrass.

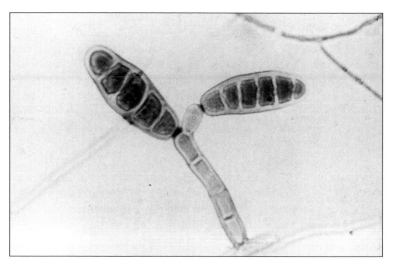

Figure 4-1. Conidiospores of melting-out of Kentucky bluegrass pathogen. *Courtesy Philip Larsen.*

periods of rainfall. Turf cut at a height of 1 inch (2.5 cm) is more susceptible to the disease than those mowed at 2 inches (5 cm). Also, as a
general rule, the severity of melting-out is greater when Kentucky bluegrass turf is growing under high nitrogen fertilization. With the advent of
warm summer months, colonization of the plants is limited to crown and
root tissue. However, if cool, wet weather develops during this time, lesions can recur on the leaves.

• **Control:**

1. Cultural Practices—In turf with a known history of melting-out, the
total amount of fertilizer used annually should be divided between fall
and spring applications. Applications of high rates of readily available nitrogen fertilizers to the turf in the spring should be avoided. If spring fertilization is practiced, the nitrogen component of the mixture should be a
slow release formulation.

When feasible, clippings should be removed. Also, since cutting
heights of less than 1.5 inches (3.8 cm) will result in an increase in the
severity of both the foliar and crown rot phases of the disease. If the use
pattern of the area permits, the turf should always be mowed at cutting
heights in the 1.5 to 2.0 inch (3.8–5.0 cm) range. If the management
program for a stand of Kentucky bluegrass with a known history of
melting-out does call for a cutting height lower than 1.5 inches (3.8
cm), the area in question should be monitored closely and a fungicide

program initiated immediately at the first indication of an outbreak of the disease.

2. Use of Fungicides—The leaf lesion phase of melting-out can be controlled by the use of chlorothalonil, iprodione, mancozeb, trifloxystrobin, azoxystrobin, or vinclozolin. For a listing of representative trade names and manufacturers of these fungicides, see Appendix Table I.

Preventive fungicidal applications should be made at 10–14 day intervals, beginning with leaf elongation (first mowing) in the spring and continuing long as weather is conducive to disease development.

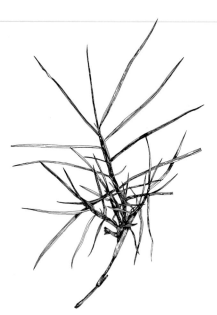

Pennisetum clandestinum
(kikuyu grass)

Helminthosporium Leaf Spot

• **Pathogen:**
Bipolaris sorokiniana (Figure 4-2)
• **Grasses Affected:**
Bentgrasses (*Agrostis* spp.), annual bluegrass (*Poa annua*), Canada blue-grass (*Poa compressa*), Kentucky bluegrass (*Poa pratensis*), tall fescue (*Festuca arundinacea*), red fescue (*Festuca rubra*), annual ryegrass (*Lolium multiflorum*), perennial ryegrass (*Lolium perenne*), bermuda-grass (*Cynodon dactylon*)
• **Season of Occurrence:**
Late spring and summer
• **Symptoms and Signs:**
1. Kentucky bluegrass—On Kentucky bluegrass, the leaf lesions are first seen as small purplish spots, and as they increase in size, the centers turn brown, and then finally fade to a light tan with purplish brown borders (Plate 4-2).

The foliar symptom pattern of Helminthosporium leaf spot is also characterized by a necrosis of the entire leaf. This blighting is manifested by a sudden collapse and drying of the leaf blades, after which the leaves blanch to a light straw color. In overall view, the disease pattern is seen as a brownish fading-out of irregularly shaped turfgrass areas of various sizes.

A severe crown and root rot frequently devel-ops in conjunction with the leaf lesion phase of the disease which appre-ciably reduces both the vigor and drought toler-ance of the plants.

2. Bentgrasses—The disease is first noticed on bentgrass golf greens and bowling greens as a smoky-blue cast of irreg-ularly shaped areas of turf, varying in size from 1 to 4 feet (0.3–1.3 m) in

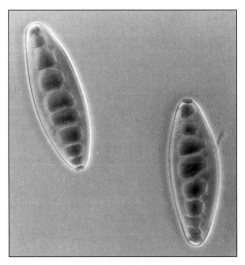

Figure 4-2. Conidiospores of Helminthospo-rium leaf spot pathogen. *Courtesy Austin Hagan.*

diameter. This symptom is soon
followed by yellowing and then
complete blighting of the leaves
within the affected areas. In advanced stages of disease development, these areas show semi-
definite margins, the leaves appear
water soaked and are matted
down.
• **Disease Profile:**
Helminthosporium leaf spot is a
warm weather disease. The first
leaf lesions usually appear in late
spring, with disease severity increasing at the onset of warm,
wet weather, and decreasing with
the advent of cooler, fall weather.
The disease is more severe on
Kentucky bluegrass that is
mowed to a height to 2 inches (5
cm) rather than 1 1/2 inches (3.8
cm), and when the turf is growing
at low soil moisture levels or high nitrogen fertilization.

Plate 4-2. Leaf spot phase of Helminthosporium leaf spot of Kentucky bluegrass.

The types of symptoms and the degree of severity of the foliar phase
of Helminthosporium leaf spot are directly related to atmospheric temperatures. At 68 °F (20 °C), leaf spotting occurs, but there is no leaf
blighting; at 75 °F (24 °C), leaf spotting predominates, with a low order
of blighting occurring; at 85 °F (30 °C), some leaf spotting occurs, but
leaf blighting is the predominant symptom; and at 95 °F (35 °C), leaf spotting is completely absent, but leaf blighting is extensive.
• **Control:**
1. Cultural Practices—In turf with a known history of Helminthosporium leaf spot, applications of high rates of nitrogen fertilizers from June
through early September should be avoided. If water soluble nitrogen fertilizers are used during this time, they should be applied on a split rate
schedule in amounts just sufficient to support an adequate plant growth
rate. The thatch layer should not be allowed to accumulate to a thickness
greater than 0.5 inch (1.3 cm).

When the use pattern of the turf permits, Kentucky bluegrass should
be mowed at cutting heights in the 1.5 to 2.0 inch (3.8–5.0 cm) range, and
if feasible, all clippings should be removed.

3. Use of Fungicides — The leaf lesion phase of Helminthosporium leaf spot can be controlled by the use chlorothalonil, iprodione, mancozeb, azoxystrobin, trifloxystrobin, or vinclozolin. For a listing of representative trade names and manufacturers of these fungicides, see Appendix Table I.

In stands of turfgrass with a history of this disease, the first treatment should be made immediately after the first occurrence of night temperatures that do not drop below 70 °F (21 °C) and the relative humidity during the nighttime period under consideration is 85 percent or higher, and continued at 10 to 14 day intervals as long as weather is conducive to disease development.

Pennisetum purpureum
(napier grass)

Red Leaf Spot of Bentgrass

- **Pathogen:**
Drechslera erythrospila (Figure 4-3)
- **Grasses Affected:**
Colonial bentgrass (*Agrostis tenuis*), creeping bentgrass (*Agrostis palustris*), velvet bentgrass (*Agrostis canina*)
- **Season of Occurrence:**
Late spring and summer
- **Symptoms and Signs:**
Individual leaf lesions are circular to oval, straw-colored, and surrounded by reddish-brown borders of variable width. On occasion, the characteristic lighter colored centers may be either extremely minute, or entirely absent (Plate 4-3). During periods of prolonged, wet weather, many of the lesions may be further surrounded by a belt of water-soaked tissue. Under conditions favorable for disease development, lesions may overlap, producing pseudozonate patterns and giving the affected areas of turf a distinct reddish cast (Plate 4-4).

Severe colonization is often accompanied by withering of the leaves beginning at the tip and progressing toward the sheath. As the result, an overall view of a diseased stand of bentgrass may have a drought-stricken appearance, even though soil moisture may be adequate for plant growth.

- **Conditions Favoring**
  Disease Development:
Red leaf spot is a warm, wet weather disease. Initial infections of crown and leaf tissue usually occur in late spring. Infection and extensive colonization of crowns, roots and tillers often occurs before the leaf lesion phase of the disease is evident. Disease incidence and severity increases as the air temperatures rise in June and July and peaks in late July and August. During periods of prolonged warm, wet weather, the leaf blighting and crown and root rot phases

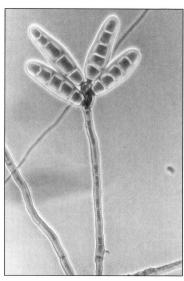

Figure 4-3. Conidiospores of red leaf spot pathogen. *Courtesy Austin Hagan.*

of the disease may become severe, leading to the death of the plants.

The severity of red leaf spot is greater in turf under high nitrogen fertilization. Also, the development of the disease is more severe when the soil is deficient in either phosphorous, potassium, or calcium.

• **Control:**

1. Cultural Practices—Care should be taken to assure that the fertilization program is maintaining a proper balance of phosphorous, potassium and calcium. If it becomes necessary to apply a water soluble nitrogenous fertilizer to the bentgrass turf during either

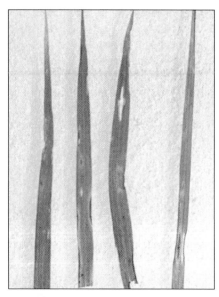

Plate 4-3. Leaf spot phase of read leaf spot of creeping bentgrass.

late spring or summer, the product should be used in amounts just sufficient to support an adequate plant growth rate. Also, the thatch layer should not be allowed to accumulate to a thickness greater than 0.5 inch (1.3 cm).

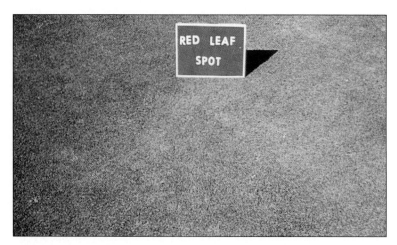

Plate 4-4. Overall view of red leaf spot of creeping bentgrass under golf green management.

2. Use of Fungicides—The leaf lesion phase of red leaf spot can be controlled by the use of chlorothalonil, iprodione, mancozeb, or vinclozolin. For a listing of representative trade names and manufacturers of these fungicides, see Appendix Table I.

A preventive fungicide program is essential for effective control of this disease. In bentgrass turf with a known history of red leaf spot, an initial fungicide application should be made at the time of spring fertilization. Successive applications should begin with the advent of warm humid weather in late spring and continued on a 7–10 day schedule as long as the climatic conditions are favorable for disease development.

Plantago lanceolata
(buckhorn plantain)

Helminthosporium Blight of Fescue, Ryegrass, and Bluegrass

• **Pathogen:**

Drechslera dictyoides (Figure 4-4)

• **Grasses Affected:**

Kentucky bluegrass (*Poa pratensis*), tall fescue (*Festuca arundinacea*), Chewings fescue (*Festuca rubra* var. *commutata*), red fescue (*Festuca rubra*), annual ryegrass (*Lolium multiflorum*), perennial ryegrass (*Lolium perenne*)

• **Season of Occurrence:**

Spring

• **Symptoms and Signs:**

On tall fescue, Kentucky bluegrass, and ryegrass the disease appears first as short, irregular, dark brown, transverse bars, which resemble strands of dark thread drawn across the leaf. These bars eventually combine with short longitudinal streaks of brown tissue, producing a very finely developed network. Under optimum conditions for disease development, these netlike patterns aggregate; fusing into dark-brown, solid lesions, measuring 1/4 to 1 inch (0.6–2.5 cm) long and 1/16 to 1/8 inch (2–3 mm) wide

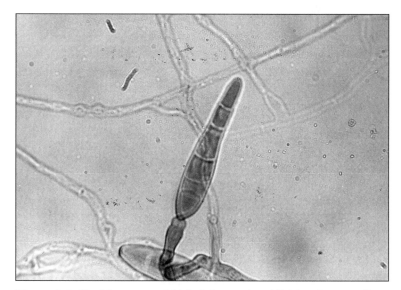

Figure 4-4. Conidiospores of Helminthosporium blight pathogen. *Courtesy Austin Hagan.*

(Plate 4-5). Heavily colonized leaves ultimately turn yellow and die back from the tips.

On creeping red fescue, the typical lesions are small, reddish-brown, irregularly shaped blotches. Leaf girdling by lesions occurs frequently, causing a yellowing and die-back from the tip (Plate 4-6).

In the warmer part of the summer, the crown rot phase of the disease causes heavily diseased stands of red fescue and tall fescue to go off color. At first they become yellow, and then finally fade to a light brown. At this time, characteristic "pockets" of dead turfgrass, ranging from 1–3 feet (0.3–0.9 m) in diameter, may develop.

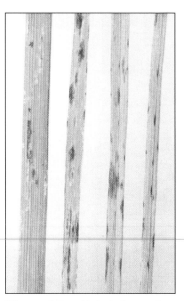

Plate 4-5. Helminthosporium blight of tall fescue.

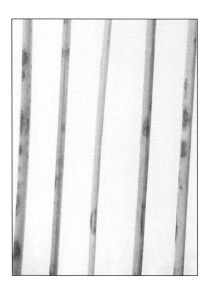

Plate 4-6. Helminthosporium blight of creeping red fescue.

• **Conditions Favoring Disease Development:**
Helminthosporium blight is a cool, wet weather disease. Infection of crowns and roots of new plants also occurs in the spring. With the advent of warmer, drier weather in late spring and early summer, the leaf lesion phase of the disease decreases and severity of the crown and root rot phase increases. During late July and August, it is not uncommon for entire stands to be rendered useless due to the crown and root rot phase of this disease. The crown and root rot phase of

Helminthosporium blight is one of the primary causes of summer "browning-up" of red fescues.

- **Control:**

1. Cultural Practices—In turf with a known history of Helminthosporium blight, the total amount of fertilizer used annually should be divided between fall and spring applications. Applications of high rates of readily available nitrogen fertilizers to the turf in the spring should be avoided. If spring fertilization is practiced, the nitrogen component of the mixture should be a slow release formulation.

2. Use of Fungicides—The leaf lesion phase of Helminthosporium blight can be controlled by the use of chlorothalonil, iprodione, trifloxystrobin, azoxystrobin, mancozeb, or vinclozolin. For a listing of representative trade names and manufacturers of these fungicides, see Appendix Table I.

Preventive fungicidal applications should be made at 10–14 day intervals, beginning with leaf elongation (first mowing) in the spring and continuing at 10–14 day intervals as long as weather is conducive to disease development.

Brown Blight of Ryegrass

- **Pathogen:**

Drechslera siccans (Figure 4-5)

- **Grasses Affected:**

Tall fescue (*Festuca arundinacea*), annual ryegrass (*Lolium multiflorum*), perennial ryegrass (*Lolium perenne*)

- **Season of Occurrence:**

Spring

- **Symptoms and Signs:**

Leaf lesions are of two types. The first are seen as small, ovular, chocolate-brown spots that eventually develop white centers. These may be very numerous, at times numbering as high as 100 per leaf. This large population of lesions gives an appearance similar to the Helminthosporium blight net blotch effect on tall fescue; however, the characteristic transverse markings of the latter disease are absent (Plate 4-7). The second lesion type takes the form of dark brown streaks 3/8 inch (6–12 mm) or more in length. Both lesion types may appear simultaneously on the same leaf blade. A high incidence of lesions usually causes the entire leaf to become blighted. This withering process begins at the tips of the leaves as a yellow discoloration and then progresses toward the sheath.

- **Conditions Favoring Disease Development:**

The leaf lesion phase of the disease develops first during the cool, wet weather of early spring. At this time, infection of crowns and roots of new plants also occurs. Secondary spread of the spores of the pathogen from plant to plant is accomplished by wind and by splashing water. During the warm summer months, the incidence of leaf lesions subsides and the severity of

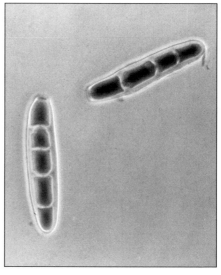

Figure 4-5. Conidiospores of ryegrass brown blight pathogen. *Courtesy Austin Hagan.*

crown and root rot increase. However, with the advent of the cool, wet weather of fall, lesions can again develop on the leaves.

• **Control:**

There is no information on the control of the leaf lesion phase of the disease.

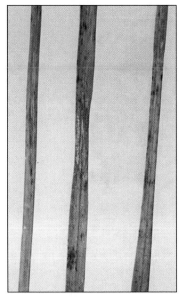

Plate 4-7. Brown blight of perennial ryegrass.

Zonate Eyespot of Bentgrass, Bluegrass and Bermudagrass

* **Pathogen**

Drechslera gigantea (Figure 4-6)

* **Grasses Affected:**

Bermudagrass (*Cynodon dactylon*), creeping bentgrass (*Agrostis palustris*), velvet bentgrass (*Agrostis canina*), Kentucky bluegrass (*Poa pratensis*)

* **Season of Occurrence:**

Summer

* **Symptoms and Signs:**

On turfgrass mowed at 1 inch (2.5 cm) and higher, zonate eyespot is first seen as minute, brown spots on the leaves. These lesions increase in length and width, finally becoming elongate, with the centers fading to dull white or light straw color. At this stage of development, the lesions are surrounded by narrow, brown borders. This difference in coloration, coupled with the elongate shape of the older lesions produces the eyespot feature that is characteristic of this disease.

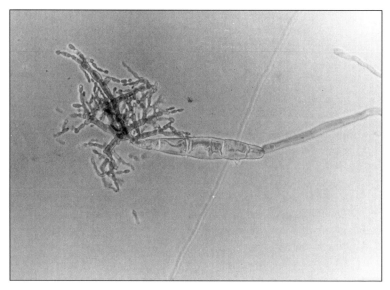

Figure 4-6. Macroconidium and microconidia of zonate eyespot pathogen. *Courtesy of Austin Hagan.*

Under conditions favorable for disease development, the lesions sometime occupy the entire width of the leaf blade. When this happens, these areas usually develop a number of irregularly concentric brown markings, giving them a distinctly zonate appearance (Plate 4-8). The complete blighting and total browning of leaves due to coalescence of lesions is not uncommon.

On bentgrass managed at golf green and bowling green mowing height, zonate eyespot is seen in overall view as irregularly shaped patches of blighted turfgrass varying in size from 2 to 9 inches (5–23 cm) in diameter. Individual leaf symptoms first develop as small yellow lesions.

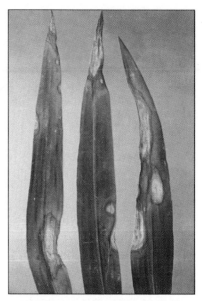

Plate 4-8. Zonate eyespot symptoms on smooth crabgrass.

These individual lesions enlarge rapidly and soon coalesce, causing the entire leaf to turn yellow, wither, and then turn brown.

• **Conditions Favoring Disease Development:**
Zonate eyespot is a warm, wet weather disease. Outbreaks usually first occur in mid to late June. Development of the disease usually reaches its peak in late August.

Highly localized outbreaks of zonate eyespot within a stand of turfgrass are of common occurrence. The spores of the pathogen usually do not survive for more than 14 days after maturity. This restricted pattern of disease development, then, is probably caused by abrupt changes in the microclimate which (a) prevents leaf infections in new turf areas by spores that have already been produced, and (b) limits the further production of spores. As the result, the disease does not spread—it stays confined to the original colonization site.

• **Control:**
There is no information on the degree of resistance of various cultivars to zonate eyespot or the effectiveness of fungicides for the control of the leaf lesion phase of this disease.

Leaf Blotch of Bermudagrass

- **Pathogen:**
Bipolaris cynodontis (Figure 4-7)
- **Grasses Affected:**
Bermudagrass (*Cynodon dactylon*)
- **Season of Occurrence:**
Spring and early fall
- **Symptoms and Signs:**
In overall view, affected areas are straw-colored, irregular in outline, and range from 2 inches (5 cm) to several feet (2–3 m) in diameter. Leaf lesions are first seen as small, olive-green spots. As these enlarge, they form irregularly shaped blotches, which are, in turn, brownish green to black.

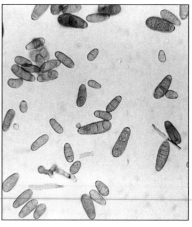

Figure 4-7. Conidiospores of bermudagrass leaf blotch pathogen. *Courtesy Austin Hagan.*

Plate 4-9. Leaf blotch of bermudagrass.

When lesions are numerous, the leaves wither and gradually fade to a light tan (Plate 4-9).
- **Conditions Favoring Disease Development:**
Development of the leaf lesion phase of the disease is favored by cool, wet weather; consequently, leaf blighting is most common during the months of late winter and early spring. With the advent of warm, relatively dry summer months, the incidence of leaf lesions decreases and the severity of crown and root rot incited by the pathogen increases. The severity of both the leaf lesion and root and crown rot phases of the disease is greater on bermudagrass growing under low potassium nutrition.

• **Control:**

The bermudagrass cultivars 'Tifgreen,' 'Turftex,' and 'Ormond' are highly resistant to Helminthosporium leaf blotch. The severity of both the leaf lesion and crown rot phases of the disease can be lessened by employing a fertilization program that provides the proper balance of nitrogen, phosphorous and potassium.

There is no information on the effectiveness of fungicides for the control of the leaf lesion phase of this disease.

Rottboellia exaltata
(Raoul grass)

Gray Leaf Spot

- **Pathogen:**

Pyricularia grisea (Figure 4-8)

- **Grasses Affected:**

Perennial ryegrass (*Lolium perenne*), annual ryegrass (*Lolium multiflorum*), St. Augustinegrass (*Stenotaphrum secundatum*), bentgrasses (*Agrostis*), annual bluegrass (*Poa annua*), Kentucky bluegrass (*Poa pratensis*), bermudagrass (*Cynodon dactylon*), centipedegrass (*Eremochloa ophiuroides*), red fescue (*Festuca rubra*), tall fescue (*Festuca arundinacea*)

- **Season of Occurrence:**

Summer

- **Symptoms and Signs:**

Leaf lesions begin as olive-green to brown water-soaked dots smaller than a pin head. These enlarge rapidly to form spots that are at first round to oval, and then elongate. The maximum size of individual lesions varies with turfgrass species. On ryegrasses, centipedegrass, and bermudagrass, older lesions usually measure 1/16 inch (1.5 mm) wide by 1/8 inch (3 mm) long, while on St. Augustinegrass, mature lesions may range up to 1/2 inch (1.3 cm) in length (Plate 4-10; Plate 4-11; Plate 4-12).

As the lesions enlarge, they develop depressed, blue-gray centers with slightly irregular purple to brown margins that are in turn bordered by a ring of yellow tissue. During extended periods of leaf wetness, the pathogen sporulates abundantly in the centers of the spots giving them a velvety gray appearance. Leaves with high numbers of lesions become yellow, then they wither and turn brown. In overall view, a severely affected turf may appear scorched as though drought stricken. (Plate 4-13).

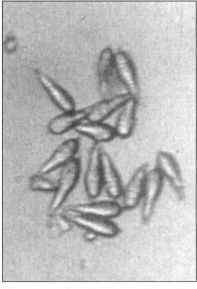

Figure 4-8. Conidiospores of gray leaf spot pathogen.

Plate 4-10. Gray leaf spot lesions on perennial ryegrass.

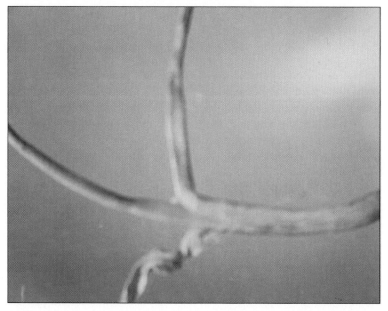

Plate 4-11. Gray leaf spot of perennial ryegrass showing twisting of leaves.

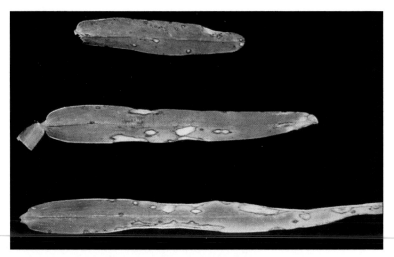

Plate 4-12. Gray leaf spot of St. Augustinegrass.

Symptoms on the spike and sheath closely resemble those on the blade, while spots on the culm are brown to black. Culm and spike infection is commonly only during periods of high disease incidence.

• **Conditions Favoring Disease Development:**

Gray leaf spot occurs during moderate to warm weather accompanied by periods of prolonged leaf wetness. Free water on the leaf surface is required for spore germination. Development of the disease in epidemic proportions requires an extended succession of continual leaf wetness periods of 24 hours or longer and air temperatures in the 70–85 °F (21–29 °C) range. Incidence of the disease on St. Augustinegrass is greater when the plants are grown under high nitrogen fertilization.

• **Control:**

1. Cultural Practices—Applications of high rates of nitrogen fertilizers during wet summer months should be avoided. If water soluble nitrogen fertilizers are used during this time, they should be applied on a split rate schedule in amounts just sufficient to support an adequate plant growth rate.

Decreasing the length of time the leaves are wet will aid in reducing the incidence of gray leaf spot. The duration of the period of daily leaf wetness can be reduced by 2 to 4 hours by following a nighttime watering schedule in which the irrigation is performed after sunset and completed before sunrise.

Plate 4-13. Overall view of gray leaf spot of perennial ryegrass.

2. Use of Resistant Grasses—The nonblue-green St. Augustinegrass cultivars 'Roselawn' and 'Florida Common' and the blue-green cultivar 'Florotam' are resistant to gray leaf spot. The blue-green cultivars 'Bitterblue' and 'Floratine' are highly susceptible to the disease.

Annual ryegrass is more susceptible to gray leaf spot than perennial ryegrass (*Lolium perenne*).

3. Use of Fungicides—Effective control of gray leaf spot can be realized with applications of either azoxystrobin, trifloxystrobin, or chlorothalonil. For a listing of representative trade names and manufacturers of these fungicides, see Appendix Table I.

Cercospora Leaf Spot
of St. Augustinegrass

- **Pathogen:**
Cersospora fusimaculans (Figure 4-9)
- **Grasses Affected:**
St. Augustinegrass (*Stenotaphrum Secundatum*)
- **Season of Occurrence:**
Late spring and summer
- **Symptoms and Signs:**
Dark brown to purple oblong spots develop on the sheath and leaf blades. Newly formed spots are uniformly colored but tend to develop a tan center as the disease develops. Lesions are usually distinct, but some coalescence does occur when disease incidence is high. Individual lesions measure 1/16 to 1/8 inch (1–3 mm) wide by 1/8 to 1/4 inch (3–6 mm) long (Figure 4-9; Plate 4-14). Severely affected leaves turn yellow and eventually die, causing the turf to have a thinned-out appearance. Sporulation of the pathogen during moist weather causes the spots to develop a whitish sheen.

At certain stages of disease development, Cercospora leaf spots closely resemble those incited by the gray leaf spot pathogen. However, when the total syndromes for the two diseases are taken into consideration, the gray leaf spot lesions are usually larger, more oval, and lighter brown than those incited by the Cercospora leaf spot fungus.

- **Conditions Favoring Disease Development:**
The development of Cercospora leaf spot is favored by warm humid weather. Outbreaks of the disease in epidemic proportions require prolonged periods of continual leaf wetness and air temperatures in the 70–80 °F (21–27

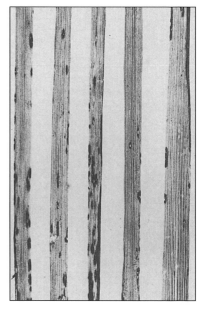

Figure 4-9. Cercospora leaf spot of tall fescue.

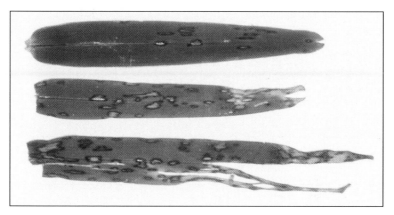

Plate 4-14. Cercospora leaf spot of St. Augustinegrass. *Courtesy T. E. Freeman.*

°C) range. Also, the severity of Cercospora leaf spot is greater when the plants are grown under low nitrogen fertilization.

• **Control:**

1. Cultural Practices—A significant reduction in disease severity can be achieved by increasing the levels of nitrogen fertilization. However, high nitrogen fertilization brings about an increase in gray leaf spot of St. Augustinegrass. Therefore, if both diseases are occurring concurrently in the same turf, then an increase in the rate of nitrogen fertilization to reduce the severity of Cercospora leaf spot will need to be balanced by a schedule of fungicide applications to compensate for the increase in severity of gray leaf spot.

Decreasing the length of time the leaves are wet will aid in reducing the incidence of Cercospora leaf spot. The duration of the period of daily leaf wetness can be reduced by 2 to 4 hours by following a nighttime watering schedule in which the irrigation is performed after sunset and completed before sunrise.

2. Use of Resistant Grasses—The yellow-green types of St. Augustinegrass are more susceptible to Cercospora leaf spot than the blue-green ones.

3. Use of Fungicides—Effective control of Cercospora leaf spot has been achieved with applications of chlorothalonil. For a listing of representative manufacturer and trade name for this fungicide, see Appendix Table I.

Notes

5

Molds, Mildews, Rusts, and Smuts

Introduction

Among the diseases in this group, powdery mildew and rusts commonly develop on turfgrasses used in inscape plantings and enclosed arboretums as well as landscape, sports, and golf turf. The slime molds are not parasites; they grow saprophytically on the surfaces of the plant. However, during periods of continuously wet weather, their development can become so extensive that it blocks out sunlight, causing the leaves to turn yellow and die. The development of downy mildew is also favored by prolonged wet weather. During times of extended rainfall, and/or standing water on the turf surface, this disease can impact severely the health of St. Augustinegrass as well as a wide range of cool season grasses.

Slime Molds

• **Pathogens:**
Mucilago spongiosa, Physarum cinereum
• **Grasses Affected:**
All commonly cultivated turfgrasses, as well as the grass weed species found in association with them, are colonized by these two species of slime molds.
• **Season of Occurrence:**
Spring, summer and early fall
• **Symptoms and Signs:**
All above-ground plant parts as well as the surface of the thatch or soil may be covered with a creamy white to translucent, slimy growth. With time, this slimy overgrowth changes to distinctive, ash-gray fruiting structures. The affected areas of turfgrass assume a dull gray appearance due to the high population of fruiting bodies on the leaves. The shape of these areas ranges from circular to serpentine streaks, and they vary in width and length from a few inches (5–10 cm) to several feet (3–5 m) (Plate 5-1; Plate 5-2; Plate 5-3).
• **Conditions Favoring Disease Development:**
Slime mold development is favored by prolonged periods of wet weather. Under cool, humid conditions, the spores of the pathogens absorb water,

Plate 5-1. Overall view of slime mold on Kentucky bluegrass turf.

Plate 5-2. Slime mold on Kentucky bluegrass leaves.

the cell walls crack open, and a single, motile swarm spore emerges from each. These motile swarm spores ingest microorganisms and decaying organic matter, and leave behind the undigested debris. With time, they divide several times and then change their form by retracting their flagella and becoming more rounded.

Ultimately, the spores unite in pairs, nuclear fusion results, and zygotes are formed. The zygotes are nonflagellate, and continue their existence as naked, ameboid cells which ingest food, increase in size, and become multinucleate due to a series of mitotic nuclear divisions. This growth form of the organism is called the **plasmodium,** and is the stage of growth that creates the slimy ct appearance on the leaves of turfgrass plants. Eventually, the plasmodium undergoes changes that lead to the extensive production of fructifications. The process by which the fruiting structures are formed is a complex one, and it varies among species. Spread of the pathogens is accomplished primarily by wind-borne spores.

With the accumulation of the plasmodium on the surface of the leaves, there is an exclusion of light and thus an interference with respiration and transpiration processes. This in turn leads to a disturbance of the metabolic activity of the underlying leaf cells. As the fruiting structures of the slime molds are formed, even more light is excluded form the leaves. In time, the impaired physiological activity of the leaf cells results

Plate 5-3. Slime mold on tall fescue leaves.

in overall leaf chlorosis. Consequently, the leaves are predisposed to invasion, and, in some cases, complete destruction by secondary bacteria, yeasts, and fungi.

• **Control:**

1. Cultural Practices—Removal of the spore masses by washing the leaves with a stream of water has been a standard recommendation for control of slime molds for many years. This method, however, should not be used during times of prolonged wet weather. If the leaves are washed during periods marked by frequent rain showers, this will only serve to spread the pathogens to previously unaffected areas, and, thus promote a buildup in disease incidence. Leaf washing for slime mold control, then, should not be performed in cases of forecasts of prevailing wet weather.

When prevailing weather conditions do not permit leaf washing, mechanical removal of spore masses raking, brushing, or poling, the affected areas will aid materially in the reduction of disease severity.

2. Use of Fungicides—Slime molds may be effectively controlled by the foliar application of any turfgrass fungicide. For a listing of fungicides for turfgrass use, see Appendix Table I.

Downy Mildew (Yellow Tuft)

• **Pathogen**

Sclerophthora macrospora

• **Grasses Affected:**

Formerly, it was thought that "downy mildew" of St. Augustinegrass (*Stenotaphrum secundatum*) and "yellow tuft" of bentgrasses (*Agrostis* spp.) were caused by different pathogens. It has now been determined the both afflictions are caused by the same organism. It has also been found that annual bluegrass (*Poa annua*), Kentucky bluegrass (*Poa pratensis*), red fescue (*Festuca rubra*), tall fescue (*Festuca arundinacea*), and perennial ryegrass (*Lolium perenne*) are susceptible to this same organism. The term "downy mildew" has been retained as the preferred name for the disease on both St. Augustinegrass and cool season grasses.

• **Season of Occurrence:**

Late winter, spring and early fall

• **Symptoms and Signs:**

On the cool season grasses, dense, yellow clusters (tufts) of plants 1 to 3 inches (3–10 cm) in diameter develop during cool wet weather in late winter and early spring and again in early fall. Each patch is made up of a dense cluster of yellow shoots that have grown from buds in the crowns and on stolons. Adventitious root development on the shoots within these clusters is very sparse, therefore the individual tufts can be easily pulled loose from the adjacent turf (Plate 5-4; Plate 5-5).

The symptoms of downy mildew on St. Augustinegrass are characterized by white, raised linear streaks that develop parallel to the midvein of the leaves. In addition to the streaking, leaves become yellow and then die back from the tips. Under conditions of high humidity, a "down" of fungal growth appears on the surface of the leaves. When drying

Plate 5-4. Downy mildew (yellow tuft) of creeping bentgrass golf green. *Courtesy Noel Jackson.*

Plate 5-5. Downy mildew (yellow tuft) of Kentucky bluegrass. *Courtesy Noel Jackson.*

of the leaf surfaces occurs, the down takes on the appearance of a dirty-white residue (Plate 5-6; 5-7).

The white streak symptom seen in the early stages of down mildew can be mistaken for the leaf-streaking-symptom associated with the

Plate 5-6. Downy mildew pathogen on Kentucky bluegrass. *Courtesy Noel Jackson.*

virus-incited disease, St. Augustine decline. However, the St. Augustine decline symptoms are more yellow, and the striping symptom pattern is not as pronounced as that of downy mildew. Also, with downy mildew, severely diseased leaves ke have a wrinkled appearance.

• **Conditions Favoring Disease Development:** Downy mildew is a cool, wet weather disease. Outbreaks are most likely to occur when the air temperatures are in the 60 to 70 °F (16–21 °C) range during periods of frequent rainfall. Turf is particularly vulnerable to severe outbreaks of downy mildew when it is growing in low-lying areas that are subjected to frequent flooding, or in locations in which water-saturated soil exists for extended periods of time.

Plate 5-7. Early symptoms of downy mildew on St. Augustinegrass leaves. *Courtesy Robert Haygood.*

The susceptibility of turfgrasses to the downy mildew fungus is not affected by soil pH or levels of fertilization. Symptoms of the disease on cool season grasses are less noticeable when either iron sulfate or nitrogenous fertilizers are applied but the effect is one of masking the symptoms rather than increasing the resistance of the plants to the disease.

• **Control:**

1. **Cultural Practices**—Since soil saturation is a necessary requirement for the development of downy mildew, good drainage of the turf area and maintenance of adequate water infiltration rates in the soil through various coring procedures are important aspects of the successful control of the disease. Iron sulfate at 10 to 20 pounds per acre (11–22 kg/ha) will mask downy mildew symptoms in sod fields.

2. **Use of Fungicides**—Control of downy mildew can be accomplished with applications of the penetrant fungicides fosetyl Al, mefenoxam or propamocarb. For a listing of representative trade names and manufacturers of these fungicides, see Appendix Table I.

Powdery Mildew

- **Pathogen:**
Erysiphe graminis
- **Grasses Affected:**
Bermudagrass (*Cynodon dactylon*), Kentucky bluegrass (*Poa pratensis*), Chewing's fescue (*Festuca rubra* var. *commutata*), red fescue (*Festuca rubra*), sheep fescue (*Festuca ovina*)
- **Season of Occurrence:**
Can occur in shaded areas during spring, summer and early fall.
- **Symptoms and Signs:**
The damage to turf by powdery mildew is usually minimal. However, the disease can be a major problem when susceptible cultivars are grown in shaded locations or areas with poor air circulation.

The primary diagnostic feature of the disease is growth of the pathogen on the surface of the leaves. It is first seen as isolated wefts of fine, gray-white, talcum powder appearing growth that is confined for the most part to the upper surface of the leaves. This growth rapidly becomes more dense, and may involve the entire leaf surface. After this, the individual leaves assume a gray-white appearance. In cases of high disease incidence, sections, or entire turfgrass stands, may be dull white, rather than the characteristic green (Plate 5-8; Plate 5-9).

Plate 5-8. Powdery mildew fungus on surface of Kentucky bluegrass leaves.

As colonization of the leaves by the pathogen becomes more intense, chlorotic lesions develop. These gradually enlarge, turning pale yellow in the process. In the final stages of disease development, the entire leaf may be pale yellow.

- **Conditions Favoring Disease Development:**

Optimum environmental conditions for the development of powdery mildew include: (a) reduced air circulation, (b) high atmospheric humidity but not visibly free water on the surfaces of the leaves, (c) low light intensity, and (d) an air temperature of 65 °F (18 °C). The importance of reduced light intensity in the development of powdery mildew is illustrated by the fact that the disease is usually more severe on turfgrass growing in shaded areas than in full natural light. This is probably due to lowered wind movement and reduced air temperatures, thus allowing the leaf surfaces to reamin wet for longer periods of time. It could, however, be the result of altered metabolism of the plant.

Plate 5-9. Powdery mildew fungus on surface of Kentucky bluegrass leaves.

- **Control:**

1. Cultural Practices—Where powdery mildew frequently recurs, if possible, the employment of management practices to improve air drainage and reduce shading will aid in the reduction of disease severity. It is not suggested, however, that the "axe be applied to the spreading chestnut tree" to control the powdery mildew on the grass below. In this case, a fungicidal program is definitely preferred as the most desirable approach.

2. Use of Fungicides—Effective control of powdery mildew can be accomplished by applications of either myclobutanil, propiconazole or triadimefon. For a listing of these fungicides and representative manufacturers, see Appendix Table I.

Rusts

- **Pathogen:**
Puccinia spp., *Uromyces* spp. (Plate 5-10)
- **Grasses Affected:**
Kentucky bluegrass (*Poa pratensis*), annual bluegrass (*Poa annua*), bent-grasses (*Agrostis* spp.), tall fescue (*Festuca arundinacea*), red fescue (*Festuca rubra*), bermudagrass (*Cynodon dactylon*), manilagrass (*Zoysia matrella*), zoysiagrass (*Zoysia japonica*), emerald zoysia (*Zoysia tenuifolia* x *Zoysia japonica*)
- **Season of Occurrence:**
The primary time of the year when rust diseases are most severe are late summer and early fall. However, they can also occur during spring, late fall and throughout the winter months if the air temperatures are moderate.
- **Symptoms and Signs:**
Early leaf lesion development is seen as light yellow flecks. As these lesions enlarge, they may become somewhat elongate, and, in cases of high incidence, show definite orientation in rows parallel with the veins of the leaves. Finally, with the rupture of the cuticle and epidermis, the lesions develop into reddish-brown pustules. As the pustules enlarge, the cuticle and epidermis that formerly covered each is pushed back to produce a

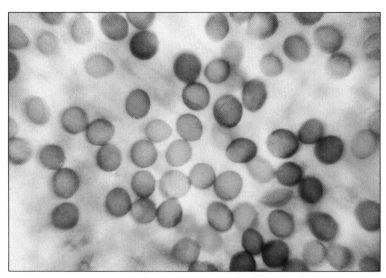

Plate 5-10. Urediospores of one of the rust pathogens, *Puccinia graminis*.

characteristic collar effect, re-
vealing yellow, red or brown
spores (Plate 5-11; Plate 5-12).

In cases of high disease incidence, the leaves of the affected
plants turn yellow, beginning at
the tips and progressing toward
the sheaths. At this stage of disease development, the entire
stand of turfgrass may appear
yellow (Plate 5-13).

• **Conditions Favoring
Disease Development:**
The development of rust is favored by a period of 2 to 3 days
of (a) overcast or cloudy weather,
(b) air temperatures in the low
70s F (21–23 °C), and (c) high
humidity brought on by fog or
frequent, light rain showers, followed by bright sunny weather
with air temperatures in the 80s F

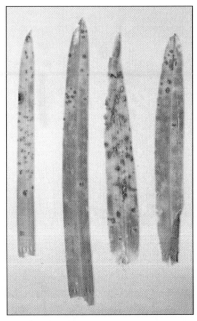

Plate 5-11. Rust on Kentucky bluegrass leaves.

Plate 5-12. Rust on perennial ryegrass leaves.

Plate 5-13. Rust damage to a highly susceptible variety of Kentucky bluegrass.

(27–30 °C). Rust is most severe on turfgrasses under low nitrogen fertilization.

• **Control:**

1. Cultural Practices—A management program that brings about a reduction in the incidence and severity of rust is one that includes collection and removal of leaf clippings, provides irrigations with short enough intervening time intervals to hold the soil in the root zone near field capacity levels (−0.033 MPa), and maintains an adequate level of nitrogen fertilization.

Management practices that decrease the length of time the leaves are wet will aid in decreasing the incidence of rust. In temperate climates, the duration of the periods of daily leaf wetness can be reduced by 2 to 4 hours by following a nighttime watering schedule in which the irrigation system is set to begin at least 3 hours after sunset and programmed to be completed before sunrise.

2. Use of Fungicides—Control of rust can be accomplished by applications of either myclobutanil, propiconazole, triadimefon, trifloxystrobin, azoxystrobin, or mancozeb. For a listing of representative trade names and manufacturers of these fungicides, see Appendix Table I.

Stripe Smut

• **Pathogen:**
Ustilago striiformis (Figure 5-1)
• **Grasses Affected:**
Bentgrasses (*Agrostis* spp.), annual bluegrass (*Poa annua*), Kentucky bluegrass (*Poa pratensis*), tall fescue (*Festuca arundinacea*), sheep fescue (*Festuca ovina*), hard fescue (*Festuca ovina* var. *brachyphylla*), perennial ryegrass (*Lolium perenne*)
• **Season of Occurrence:**
Major symptoms are most evident during late winter and early spring.
• **Symptoms and Signs:**
Turfgrass plants colonized by the stripe smut fungus usually make slow vegetative growth. Long, yellow-green streaks develop on the leaves of the affected plants, and as the disease progresses, these streaks become gray. In the final stages of disease development, the cuticle and epidermal cells covering these streaks are ruptured, exposing the underlying, black spore masses of the pathogen. After this, the leaves split into ribbons and curl from the tips downward. The leaf blades then turn light brown, wither, and die (Plate 5-14); Plate 5-15; Plate 5-16).
• **Conditions Favoring Disease Development:**
The stripe smut pathogen is both seed- and soil-borne. The spores of the fungus begin growth under soil conditions favorable for the development of turfgrass seedlings. Infection of young turfgrass seedlings occurs through coleoptiles. With older plants, tillers serve as the chief avenues of entry of the pathogen. After penetration has been accomplished, the fungus grows systemically throughout the host, and it persists in the tissues throughout the life of the plant.

The pronounced striping and shredding of the leaves, however, is brought on by periods of cool air temperatures and bright, sunny days. Plant that have been growing at 90 °F (32 °C) usually do not show these symptoms. On

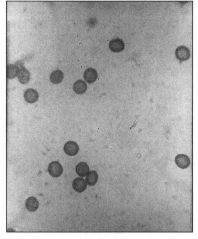

Figure 5-1. Teliospores of stripe smut fungus.

Plate 5-14. Stripe smut of Kentucky bluegrass showing yellow leaf symptom.

the other hand, when air temperatures are in the 50 to 60 °F (10–16 °C) striping and shredding of leaves can be very severe.

Spring applications of high rates of nitrogenous fertilizers can bring about a remission of stripe smut symptoms; however, in the long run this practice actually causes an increase in the intensity of the disease. The incidence and severity of stripe smut will be lowest in turf receiving adequate rates of balanced fertilizer, and highest in turf that is growing under an imbalance of either nitrogen, phosphorous or potassium.

• **Control:**

1. Cultural Practices—Some of the damage caused by stripe smut

Plate 5-15. Stripe smut of Kentucky bluegrass. Note leaf striping and leaf bending.

Plate 5-16. Stripe smut of Kentucky bluegrass, showing extensive leaf striping and leaf splitting.

can be offset by fertilization practices that promote good growth of the affected turf. Applications of high rates of nitrogenous fertilizers should be avoided. Instead, the fertilizer should be a balanced formulation and applied at intervals that provide a uniform rate of growth throughout the season. Stripe smut diseased turf is highly vulnerable to damage from heat and drought stress, therefore, during hot dry weather it is important that particular attention be given to addressing the increased irrigation needs of the affected area.

2. Use of Fungicides—Satisfactory control of stripe smut can be accomplished only by applications of penetrant fungicides. For optimum results, either propiconazole, triadimefon, fenarimol, or myclobutanil should be applied in October or in early spring before grass growth begins. A single fungicide application during one of these time periods will usually suppress the disease during the following growing season. However, if fungicidal treatment is initiated during the growing season, repeated applications may be required in order to maintain a satisfactory level of stripe smut control.

For a listing of representative trade names and manufacturers of these fungicides, see Appendix Table I.

Notes

6

Senectopathic Disorders

Introduction

Leaves of turfgrass plants follow set patterns of physiological development as they progress from juvenility to maturity, and finally into advanced stages of senescence.[1] As they become older, significant changes occur in the functions and composition of the leaves. The various metabolic activities of the cells move into a declining mode, their chemical profile changes, the rate of photosynthesis decreases, and the leaf cuticle becomes thinner. Death of the leaf is imminent.

The majority of microorganisms that are pathogenic to turfgrasses are capable of infecting and colonizing both juvenile and mature plant tissue. However, certain microbes are only able to parasitize senescent tissue. These species are referred to as **senectophytes.** Biotically incited diseases that can only develop after plant tissue is in advanced senescence are known as **senectopathic disorders.**

Within the normal growth pattern of the turfgrass plant, death of leaves due to senescence does not have a negative impact on turf quality. The reason is that although the growth processes of the older leaves are decreasing, each plant still has several younger leaves that are in active stages of growth. The presence of these younger leaves compensates for any negative effect death of the older leaves might have had on either the total well being of the plant or the overall appearance of the turf. However, certain external stress factors can cause a high population of the younger leaves in a stand of

[1]Senescence from *senescens* (L) meaning to grow old; aging.

turfgrass to shift simultaneously into advanced senescence. When this happens, a vast majority of the leaves of each plant can become infected and colonized by senectophytes, which in turn accelerates leaf death and thus a deterioration in turf quality.

Factors which can induce advanced senescence prematurely in the leaves and crowns of turfgrass plants include (a) the pathogenic effects of parasitic fungi, insects, and root feeding nematodes, (b) high air temperatures, (c) high soil temperatures, (d) low light intensity, (e) high soil moisture stress, (f) anaerobic soil conditions, (g) nutrient excesses, (h) nutrient deficiencies, (i) certain mowing and thatch management practices, and (j) toxic side effects of pesticides.

Determining whether the condition at hand is a senectopathic disorder is essential to the development of an effective control program. If the diagnostic workup has revealed the presence of a microorganism that is a known pathogen of juvenile and mature plant tissue, and the climatic conditions favor development of the disease it incites, then the application of pesticides for its control is clearly in order. On the other hand, advanced senescence is an irreversible condition that in itself soon leads to death of the affected tissue; therefore, if the disease in question is diagnosed as senectopathic, attempts to control it with pesticides will at best only be marginally successful. In this instance, for production of quality turf to be realized, the factors that are causing a large percentage of the younger leaves to become senescent simultaneously must be dealt with.

At present, the pathology of three senectopathic disorders of turfgrasses has been worked out: anthracnose, Curvularia blight, and Leptosphaerulina leaf blight.

Anthracnose

- **Pathogen:**

Colletotrichum graminicola (Figure 6-1)

- **Grasses Affected:**

Anthracnose occurs on a wide range of cool season and warm season turfgrasses. Severe cases of the disorder can develop on annual bluegrass (*Poa annua*), the bentgrasses (*Agrostis* spp.), perennial ryegrass (*Lolium perenne*), and red fescue (*Festuca rubra*) when the plants are undergoing physiological stress brought on by adverse growing conditions.

- **Season of Occurrence:**

Late winter, spring and summer

- **Symptoms and Signs:**

During warm, wet weather the older leaves of annual bluegrass turn yellow and then become tan to brown. Leaf discoloration usually starts at the tip and progresses downward to the sheath; however, under conditions of acute high air temperature stress, the entire leaf yellows simultaneously. Numerous, small, raised, black fruiting bodies (acervuli) with spines (setae) protruding from them can be seen on the surfaces of dead leaves with the aid of a magnifying lens (Plate 6-1).

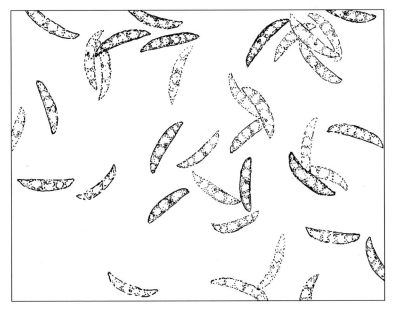

Figure 6-1. Conidiospores of anthracnose pathogen.

Plate 6-1. Acervuli with setae of anthracnose pathogen on annual bluegrass leaf.

In overall view, the annual bluegrass component of stands comprised of mixtures of turfgrass species develops a yellow-green cast which soon fades to yellow, and then becomes brown, giving the turf a blotched appearance. In dense stands of annual bluegrass, the discoloration and dying out of the turf is more uniform (Plate 6-2).

On annual bluegrass, a dark brown discoloration sometimes develops on the bases of leaf sheaths and stems. The leaves then turn yellow, beginning at the tips and progressing down the blades to the sheaths. The older leaves are the first to be affected. The sheath and stem bases eventually turn black and the stems can be easily pulled loose from the crowns (Plate 6-3). Irregular patches ranging from 1/2 to 6 inches (1–15 cm) in diameter may develop in dense stands of annual bluegrass. These areas are at first yellow, then brick red, and finally brown. On bentgrasses, there is a blackening of the sheath and crown tissues that at times extends into the adventitious roots. The older leaves are affected first. Scattered patches may form that are irregular in outline, and range in diameter from 1/2 to 18 inches (1–45 cm) or more. Their color is first gray green to tan and finally a dull brown.

•   **Conditions Favoring Disease Development:**
The basal rot form of anthracnose on annual bluegrass and bentgrasses develops when the air temperatures range between 60 and 75 °F (15–24

Plate 6-2. Overall view of anthracnose of annual bluegrass component of creeping bentgrass golf green. *Courtesy Ned Tisserat.*

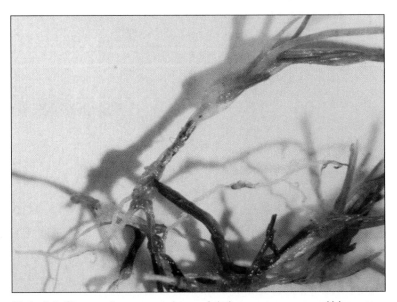

Plate 6-3. Stem and crown rot phase of anthracnose on annual bluegrass. *Courtesy Ned Tisserat.*

°C) and subsides with the advent of warmer weather. The leaf blight stage of the disease occurs on annual bluegrass, bentgrasses, red fescue, and perennial ryegrass after there has been a succession of daytime air temperatures between 85 and 95 °F (29–35 °C) combined with extended periods of leaf wetness.

Infection and colonization of mature turfgrass plants by the anthracnose fungus occurs only when the tissue in question has been weakened by biotic and abiotic stress factors such as (a) high populations of root, crown or leaf-feeding insects, (b) high populations of root-feeding nematodes, (c) parasitism of the crown and leaf tissues during their juvenile and mature stages of growth by pathogenic fungi, (d) extremes in soil fertility levels, (e) high air temperatures, (f) cold injury—including crown hydration—(g) anaerobic soil conditions, (h) soil moisture stress, and (i) soil compaction.

• **Control:**
Refer to the section Control of Senectopathic Disorders at the end of this chapter.

Curvularia Blight

- **The Pathogens:**
Curvularia geniculata, C. inaequalis, C. intermedia, C. lunata, C. protuberata, C. trifolii (Figure 6-2).
- **Grasses Affected:**
Annual bluegrass (*Poa annua*), Kentucky bluegrass (*Poa pratensis*), creeping bentgrass (*Agrostis palustris*), velvet bentgrass (*Agrostis canina*), red fescue (*Festuca rubra*), tall fescue (*Festuca arundinacea*), Chewing's fescue (*Agrostis rubra* var. *commutata*), zoysiagrass (*Zoysia japonica*), bermudagrass (*Cynodon dactylon*)
- **Season of Occurrence:**
Summer
- **Symptoms and Signs:**
Turf exposed to full sun and growing near sidewalks and paved areas is most severely affected by Curvularia blight. The disease is first seen in overall view as irregularly shaped, mottled patches that fade to yellow, and then become brown. Individual leaf symptoms develop as a dappled, yellow-green discoloration beginning at the tips. As the discoloration extends downward toward the leaf sheath, it progressively turns from yellow to brown. In the advanced stages of disease development, the affected leaves shrivel and die.

When conditions are particularly favorable for the development of Curvularia blight, colonization of the leaf sheaths and crowns can also occur. When the disease goes into the crown rot phase on bentgrass being managed at golf green and bowling green mowing heights, discrete tan to brown patches of blighted grass 2 to 4 inches (5–10 cm) in diameter often develop. With Kentucky bluegrass and red fescue, the crown rot phase of the disease gives the turf a ragged or thinned-out appearance.

- **Conditions Favoring Disease Development:**
The optimum conditions for outbreaks of the disease on cool season

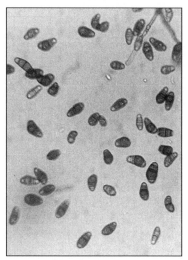

Figure 6-2. Conidiospores of one of the Curvularia blight pathogens, *Curvularia lunata*.

grasses are several days in succession of daytime air temperatures between 85–95 °F (30–35 °C) combined with extended periods of leaf wetness. With the warm season grasses, Curvularia blight develops during early spring on plants that have received winter damage and during summer and early fall on turf that has been stressed by high populations of root feeding nematodes, insects, fertility imbalances, or drought.

On creeping bentgrass, the fungus grows through cut tips or surface wounds into underlying tissue at any stage of leaf growth and remains between the cells. The actual penetration and colonization of cells does not occur until prolonged high temperature stress has caused the tissue to shift into advanced senescence. Entry of leaf or crown tissue can also take place through lesions incited by the Helminthosporium leaf spot pathogen. When this happens, regardless of air temperature, tissue colonization by the Curvularia blight fungus begins immediately. However, when air temperatures are in the 85–95 °F (30–35 °C) range, there is a synergistic effect between the Helminthosporium blight pathogen and the invading Curvularia species that results in greater disease severity than that brought on by either pathogen individually.

Spring applications of the herbicides 2,4-D (dichlorophenoxyacetic acid), MSMA (monosodium acid methanearsonate), or dicamba (3,6-dichloro-o-anisic acid) at high rates will increase the severity of Curvularia blight on creeping bentgrass during the months of high temperature stress.

• **Control:**
Refer to the section Control of Senectopathic Disorders at the end of this chapter.

Leptosphaerulina Leaf Blight

- **The Pathogens:**
Leptosphaerulina trifolii, L. australis
- **Grasses Affected:**
Annual bluegrass (*Poa annua*), Kentucky bluegrass (*Poa pratensis*), bentgrasses (*Agrostis* spp.), red fescue (*Festuca rubra*), tall fescue (*Festuca arundinacea*), annual ryegrass (*Lolium multiflorum*), perennial ryegrass (*Lolium perenne*)
- **Season of Occurrence:**
Summer
- **Symptoms and Signs:**
Individual leaf symptoms develop as a yellow discoloration beginning at the tips. As the discoloration extends downward toward the leaf sheath, the leaf progressively turns yellow and then brown. Necrosis may extend into the sheath. In the advanced stages of disease development, the affected leaves shrivel and die. Small brown fruiting structures of the fungus often develop on the upper and lower surfaces of dead leaves.
- **Conditions Favoring Disease Development:**
Both species of *Leptosphaerulina* grow saprophytically on thatch and leaf litter. Infection and colonization of leaves takes place only after the tissue has entered into advanced senescence through natural growth processes or as the result of environmental stresses.

Control of Senectopathic Disorders

1. Cultural Practices—Various cultural practices to minimize the occurrence of factors that accelerate leaf senescence will reduce both the incidence and severity of senectopathic disorders. Specifically:

- The thatch layer should not be allowed to accumulate to a thickness greater than 0.5 inch (1.3 cm).
- Turf should be syringed or lightly watered at the hottest time of day.
- Soil compaction should be reduced by coring and the employment of the most appropriate procedure to improve water infiltration rates.
- Refrain from frequent, shallow waterings. Instead, irrigations should be of sufficient duration to wet the soil throughout the rooting zone to field capacity (-0.033 MPa).
- Employ a fertilization program that prevents the development of phosphorous and potassium deficiencies.
- With the warm season grasses, avoid late summer fertilizations that may increase the plants' vulnerability to winter damage.

2. Use of Fungicides—Since senectopathic fungi can only parasitize plant tissue in an advanced stage of senescence, the economic benefit of fungicide is open to question. No commercially available fungicides carry registration for control of Curvularia blight or Leptosphaerulina leaf blight. Azoxystrobin, trifloxystrobin, thiophanate methyl, triadimefon, fenarimol, propiconazole, and triadimefon, are generally labeled for the control of anthracnose. To realize any benefit from fungicide use to reduce the incidence of anthracnose, the material must be applied on a preventive schedule. A listing of representative trade names and manufacturers of these fungicides, see Appendix Table I.

The most appropriate use of pesticides for senectopathic disorders is to control microbes capable of infecting and colonizing juvenile and/or mature plant tissue. A complete diagnostic workup should be performed and a preventive pesticide program implemented to control associated primary pathogens (fungi, nematodes, insects) that have been identified. Successful control of the diseases these organisms incite will lower the possibility of the simultaneous shift of large numbers of leaves and crowns into advanced senescence and thus significantly reduce the opportunity for either the anthracnose, Curvularia blight, or Leptosphaerulina leaf blight fungi to parasitize them.

7

Root Diseases

Introduction

The initial impact of this group of diseases is on the underground organs of turfgrasss plants—roots, stolons, and rhizomes. In overall view, turfgrass affected by root diseases shows various shades of light green to yellow, usually accompanied by an overall decline of the plants. The damaged areas will vary in shape and size, and the boundaries between healthy and diseased areas of turf are not sharply defined.

In advanced stages of disease development, the feeder root systems become thinned out and the remaining roots are often shortened and decomposed. Stolons and rhizomes develop necrotic lesions and also may become rotted. The affected plants lack vigor, and have a diminished ability to withstand dry soil conditions, low fertility, extremely high air temperatures, or reduced mowing heights.

An important aspect of successful, long term control of these diseases is the diligent use of management practices that reduce plant stress and foster the growth and development of the root systems.

Anaerobiosis (Black Layer)

Anaerobiosis is a dynamic series of events taking place in an oxygen depleted (anaerobic) environment. When the soil becomes anaerobic, there are significant changes in both the form and solubility of nutrient elements. In their reduced state, certain of these elements are taken up by the plant more rapidly than they can be metabolized, thereby becoming toxic. Others become limited in availability. Also, oxygen-depleted soils promote the growth of anaerobic microorganisms which produce metabolites that are detrimental to plant growth. Root systems become dysfunctional in anaerobic soils. Their ability to absorb water and nutrients is reduced significantly.

• **Diagnostic Features:**

There is a gradual reduction in the vigor of the turf. The plants develop a pale green to yellow color. In lower areas of a golf green subjected to a high frequency of foot and machinery traffic, there is a thinning, and finally bronzing and death of the turf in patches and strips. Root development in the affected areas is severely restricted (Plate 7-1).

In the advanced stages of anaerobiosis, black layers may develop in the soil profile. These layers may be confined to the top 1 to 2 inches (2.5–5.0 cm) of the soil, but can occur at depths of 4 to 6 inches (10–15

Plate 7-1. Golf green showing symptoms of anaerobiosis. *Courtesy Jonathan Scott.*

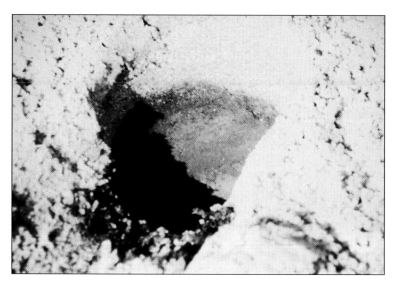

Plate 7-2. Golf sand bunker showing black layer stage of anaerobiosis. *Courtesy Jonathan Scott.*

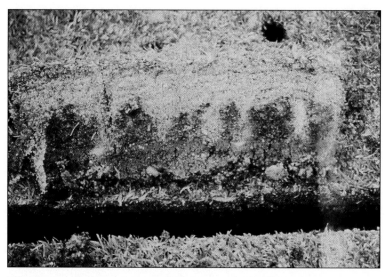

Plate 7-3. Anaerobiosis showing the impact of aerification on breakdown of the black layer. *Courtesy Jonathan Scott.*

cm) or more (Plate 7-2; Plate 7-3). When the soil is water soaked, it may develop a distinctive, "rotten egg" smell. This odor is almost unnoticeable when the black layer is dry. Over time, infiltration of water through the layer is progressively restricted.

• **Conditions Favoring Disease Development:**
Anaerobiosis is the consequence of a water-saturated soil that has been brought on by frequent flooding, extended periods of high rainfall, excessive irrigation, or inadequate drainage due either to compaction, the construction of rooting zones on inadequately drained bases, or the formation of impermeable layers within established seedbed mixes. Water-impermeable layers within seedbed mixes of sand-based golf greens can develop as the result of improper particle size distribution due to the use of incompatible top dressing, and/or the formation of a gelatinous substance (biofilm) by soil-borne anaerobic bacteria that plugs both the capillary and noncapillary pores. Sealing of the soil surface by either dust, and/or the growth of blue-green algae also fosters the development of anaerobiosis. Also, poor quality irrigation water adds daily organic fractions to the soil column.

Sulfur is a major nutrient for certain biofilm-forming bacteria, however, rooting mixtures already contain adequate sulfur to sustain abundant growth of these species. The application of elemental sulfur to the turf, the use of sulfur-coated fertilizers, or reliance on irrigation water high in sulfate neither causes anaerobiosis nor exacerbates an existing anaerobic condition.

• **Control:**
There is a direct relationship between rate of water infiltration and biofilm formation. A key element in the effective management of anaerobiosis is frequent monitoring of water infiltration rates. Ideally, the rooting mixture of sand-based bentgrass golf greens should not be allowed to drop below 5 to 7 inches (13–18 cm) per hour. The infiltration rate of the rooting mixture of sand-based bermudagrass greens should not be less than 4 to 5 inches (10–13 cm) per hour.

Measurement of the rate of water penetration of the thatch and the velocity of its movement through the underlying root zone mixture can be accomplished with a field infiltrometer (available from Turf-Tec International, 4740 NE 12th Avenue, Oakland Park, Florida 33334). The baseline (initial) infiltration rates for preselected locations on each green should be recorded. Measurements should then made periodically at these sites throughout the growing season. At the first indication of a decrease, even though it may not appear to be significant, the cause should be identified and deep tyne aerification performed to improve infiltration rates.

Pythium Root Dysfunction
of Creeping Bentgrass

- **Pathogens:**

Pythium arrhenomanes, P. aristosporum

- **Grasses Affected:**

Creeping bentgrass (*Agrostis palustris*), Kentucky bluegrass (*Poa pratensis*), annual bluegrass (*Poa annua*), tall fescue (*Festuca arundinacea*), red fescue (*Festuca rubra*), perennial ryegrass (*Lolium perenne*), St. Augustinegrass (*Stenotaphrum secundatum*)

- **Season of Occurrence:**

Summer

- **Symptoms and Signs:**

On bentgrass golf greens, the above ground symptoms of Pythium root dysfunction are first seen as small, irregularly shaped pale green to light yellow areas ranging from 1 to 4 inches (2.5–10 cm) in diameter. Pale green to light yellow strips may also develop at the interface of the high sand content mix of the golf green and the collar apron of soil. Within a few days from the first appearance of symptoms, the color of the affected areas changes to light brown. As the disease progresses, individual patches of affected grass frequently coalesce to envelop sections of turf ranging from 1 to 10 feet (0.3–3 m) in diameter. During hot, humid weather, entire greens may die within 7 to 14 days from the onset of symptoms. Certain aspects of the above ground symptom pattern for Pythium root dysfunction are similar to Pythium foliar blight, however, with Pythium root dysfunction, the leaves of the affected plants are free of colonization by Pythium species.

Examination of the root systems of diseased plants reveals white, normal appearing roots. Although there is no necrosis, when affected roots are incubated in the laboratory using tissue culture chambers, the pathogen grows from the tissues within 6–12 hours. In the advanced stages of disease development, affected roots may become slightly buff colored and develop bulbous root tips. Also, the root tips eventually become disorganized and devitalized (Figure 7-1).

- **Conditions Favoring Disease Development:**

Root dysfunction is the inability of what would appear to be a healthy root system to adequately absorb water and mineral nutrient elements from the soil. The fungi that cause this disease infect and thoroughly colonize the roots during spring and early summer but fail to produce root rot. Under optimum growing conditions, there is no evidence of adverse

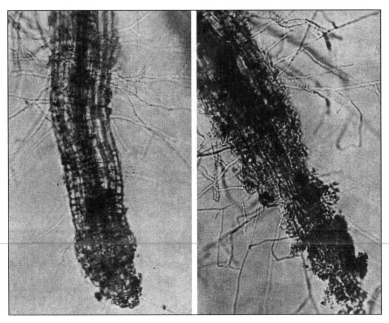

Figure 7-1. Symptoms of creeping bentgrass roots colonized with the Pythium root dysfunction pathogens. Left: Bulbous root tip and growth of mycelium from region of colonization. Right: Devitalized root tip. *Courtesy C. F. Hodges. Reprinted by permission of* Plant Disease *69:33–340.*

effects on the plants. However, when the plants are placed under stress, death of large areas of turf can occur within a relatively short period of time.

The foliar death phase of Pythium root dysfunction occurs during hot, humid weather the first or second growing season after the green has been established. Severe damage may persist for up to 3 successive years. After this, the disease usually decreases in severity, and within 5 years it either ceases to be a problem or occurs at a much reduced level of activity.

Although Pythium root dysfunction occurs commonly in sand-based golf greens, pathogenicity tests have failed to note any significant difference in severity of the disease between plants grown in either sand or sand-loam media.

•   **Control:**

1. Cultural Practices—In certain situations, coring and filling the holes with sand topdressing in order to stimulate root growth will provide some degree of recovery from the problem.

2. Chemical Control—Attempts to control Pythium root dysfunction with either contact or penetrant fungicides that are active against Pythium species have not been successful.

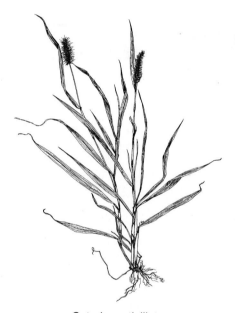

Setaria verticillata
(bristly foxtail)

Bermudagrass Decline

*   **The Pathogen:**
Gaeumannomyces graminis var. *graminis*
*   **Grasses Affected:**
Bermudagrass (*Cynodon dactylon*), hybrid bermudagrass (*Cynodon dactylon* X *Cynodon transvaalensis*)
*   **Season of Occurrence:**
Late spring and summer
*   **Symptoms and Signs:**
In overall view, the disease is first seen as irregularly shaped light yellow patches measuring from 8 inches to 3 feet (0.2–1.0 m) in diameter. Yellowing and necrosis are first observed on the lower leaves. Foliar lesions are absent. The root systems of affected plants are short and discolored, with dark-colored lesions on the roots. The surfaces of affected roots and stolons may bear dark strands of mycelium mostly running parallel to the main axes. Eventually, the roots and associated stolons become completely rotted. Entire plants may die, resulting in a thinning of the turf. In the advanced stages of disease development, bare patches may develop and coalesce. Initially, newly installed bermudagrass tillers and transplanted sod may develop in the patch areas, but they soon become colonized by the pathogen and also decline. The outer margins of a golf green often exhibit the disease symptoms first, eventually, however, the symptoms may be expressed across the entire green (Plate 7-4).
*   **Conditions Favoring Disease Development:**
The disease occurs when the continuing daily air temperature average is greater than 80 °F (27 °C), the ongoing relative humidity is higher than 75 percent, and precipitation occurs almost daily. The severity of the disease is also greater in turf maintained at cutting heights of 3/16 inch (4.8 mm) or less, or growing in soil low in potassium.

Local distribution of the pathogen from one site to the next is by transport of infested soil and diseased plant tissue on coring and dethatching equipment. Long-range transport of the pathogen and its introduction into new locations can be accomplished by the use of diseased sprigs. Since sprigs are normally planted into fumigated soil, populations of suppressive soil-borne organisms are low, which enables the pathogen to spread rapidly on the new root systems.
*   **Control:**
1. Cultural Practices—Mowing height should be maintained at no less than 1/4 inch (6.4 mm). Thatch should be kept at 1/2 inch (1.3 cm) in thickness. Coring should be performed in the diseased areas to improve

Plate 7-4. Bermudagrass decline on 'Tifgreen' bermudagrass under golf management. Left: Maintained at 3/16 inch (0.9 cm.) cutting height for 4 months. Right: Maintained at 1/4 inch (1.1 cm.) cutting height for 4 months. *Courtesy Monica Elliott.*

soil aeration and foster the production of new roots. Aerification should be accomplished to a depth of 6 inches (15 cm) with large tines and the procedure should be performed every 3 to 4 weeks during the late spring, summer, and early fall months. Cores should be removed and the area topdressed with a topsoil mix containing up to 30 percent organic matter.

The nitrogen component of the fertilization program should based on the use of acidifying fertilizers such as ammonium sulfate. Avoid nitrate nitrogen sources. Also, attention should be given to maintaining adequate potassium levels. Nitrogen and potassium should be applied in a 1:1 ratio at 1/2 to 1 pound (227–453 g) per 1000 square feet (93 m^2) per week.

2. Use of Resistant Grasses—Both 'Tifdwarf' and 'Tifgreen' cultivars of hybrid bermudagrass (*Cynodon dactylon* $\times$ *Cynodon transvaalensis*), the most widely grown cultivars on golf greens, are very susceptible to this disease.

3. Use of Fungicides—To date, attempts to control bermudagrass decline with either contact or penetrant fungicides have not been successful.

Take-all Root Rot of St. Augustinegrass

- **Pathogen:**

Gaeumannomyces graminis var. *graminis*

- **Grasses Affected:**

St. Augustinegrass (*Stenotaphrum secundatum*)

- **Season of Occurrence:**

Late spring and summer

- **Symptoms and Signs:**

Above-ground symptoms consist of chlorotic, thinning turf in circular to irregular patches 2 to 15 feet (0.6–5.0 m) or more in diameter. The over-all color of these patches is at first yellow, then brown, and finally black. Death of plants in the affected areas is common. At times, the overall appearance of the affected turf is similar to that of Rhizoctonia blight. However, unlike Rhizoctonia blight, there is no basal leaf and sheath rot, and therefore, the leaves do not separate easily from the plant. The roots of plants in these patches are short and rotted. The stolons can be readily lifted from the ground. Sod producers are unable to harvest sod because its strength is severely weakened by the root rot. Nodes are usually rotted also and black lesions develop on stolons (Figure 7-2).

- **Conditions Favoring Disease Development:**

There does not appear to be a relationship between take-all root rot incidence and severity and soil type, cultivar, or age of St. Augustinegrass. Development of the disease is favored by warm, humid weather, with major symptom expressions occurring during the summer and fall months and the frequency of daily precipitation is high. The pathogen colonizes all underground plant parts, and since St. Augustinegrass is vegetatively propagated, it is likely that the disease can be spread to new locations with sprigs or sod.

- **Control:**

1. Use of Resistant Grasses—There is no indication of varietal resistance to take-all root rot. The disease has been observed on 'Floratam,' 'Jade,' 'Raleigh' in Texas and Florida, and on common St. Augustinegrass in Alabama. Also, the pathogen has been isolated from 'Del Mar,' 'Jade,' 'Dalsa 8401,' 'Mercedes,' 'Bitterblue,' 'Standard,' California common, Scott's '138,' '770' and '2090,' 'Sunclipse' 'Raliegh,' 'Milberger M1,' 'Seville,' and 'Floratam' in California.

2. Cultural Practices—There are no fungicides registered for control of this disease; therefore, the use of acidifying fertilizers such as ammonium sulfate and maintaining adequate soil potassium levels offer the best possibility for reducing disease severity.

Figure 7-2. Severe patch symptoms associated with take-all root rot of St. Augustinegrass. Note bare ground and rotting stolons. *Courtesy Monica Elliott.*

Nematode Damage

- **Pathogens:**
Several species of root-feeding nematodes
- **Grasses Affected:**
All cool season and warm season turfgrasses
- **Season of Occurrence:**
Late spring and summer
- **Symptoms and Signs:**
The affected turf displays various shades of light green to yellow. The areas showing damage will vary in shape and size, and the boundaries between healthy and diseased turf are not sharply defined. The affected plants lack vigor, and have a reduced ability to withstand dry soil conditions, low fertility, extremely high air temperatures, or reduced mowing heights (Plate 7-5; Plate 7-6).

A credible diagnosis of root-feeding nematode damage can only be made by performing an assay on soil samples from the affected site. The purpose of the assay is to determine whether parasitic species of nematodes are present, and whether the population levels are high enough to cause the problem.

The collection of soil samples for nematode assay should not begin until the soil temperature at the 2-inch (5 cm) depth reaches 50 °F (10 °C).

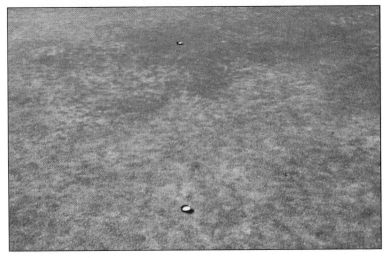

Plate 7-5. Lance and sting nematode damage to bentgrass under golf green management. *Courtesy Robert Wick.*

Plate 7-6. Sting nematode damage to bermudagrass. *Courtesy Robert Haygood.*

Using a standard 1-inch (2.5 cm) diameter soil tube, put together a composite sample of 15 to 30 cores per 500 to 1,000 square feet (46–93 m^2). Samples should be taken to a depth of 4 inches (10 cm) not including the thatch. The total volume of the composite sample for each location should be at least 1 pint (500 cc), which is approximately 20 soil cores taken to a 4 inch (10 cm) depth. The sample should be placed in an airtight bag and delivered to the nematode assay laboratory within 3 days. The minimum number of nematodes required to produce significant damage to the turf is called the **damage threshold level.** The damage threshold of the 13 major groups of root-feeding nematodes for cool season and warm season grasses respectively is listed in the third edition of *Diseases of Turfgrasses* by H. B. Couch (Krieger, 1995).

• **Conditions Favoring Disease Development:**
Plant parasitic nematodes feed only on living host cells. Penetration of the root tissue is accomplished by a syringelike structure known as the stylet. The stylet serves to inject digestive enzymes into the host cell and then withdraw the partially digested contents.

The life cycles of root-feeding nematodes are fairly simple. The females lay eggs which hatch into juveniles. The juveniles may be similar to adults in overall appearance and undergo a series of four molts before

adulthood is reached. The rate of nematode activity, growth and reproduction increases as the soil temperature rises from 50 °F (10 °C) to about
90 °F (32 °C). The minimum life span for a single generation of nematodes is about 4 weeks.

In order for root-feeding nematodes to have an adverse effect on turf
quality, within a relatively short period of time they must be able to produce population levels in sufficient numbers to bring about a high incidence of feeding sites. Certain species are more pathogenic than others;
therefore, they will cause damage at lower population levels. Also, with
any given nematode species, the population levels required to produce
varying degrees of injury to the plants is directly related to the degree of
stress to which the plants are being subjected.

Nematode population densities generally increase during the summer months. However, because parasitic nematodes can only feed on living roots, their population levels are often inversely related to the degree
of damage to turfgrass roots. Because of this, population densities are frequently higher in adjacent healthier appearing turf than in areas of poor
turf quality.

•   **Control:**
1. Cultural Practices—Reduction in the negative effect of root-feeding
nematodes on turf quality can be achieved by the use of management
practices that foster root growth and development. Irrigations should be
of sufficient duration to facilitate deep penetration of the rooting medium
rather than shallow daily waterings. The turf should be mowed at the
maximum height landscape or use conditions permit. A well-planned and
properly executed tillage program (eg., coring, aerification, spiking)
should be set in place to maintain adequate oxygen levels in the root zone
and facilitate proper water infiltration rates.
2. Use of Nematicides—Preplant fumigants can be used for nematode
control prior to the installation of the turf. On established turf under golf
course management, the organophosphate nematicide fenamiphos can be
used to reduce turf damage caused by root-feeding nematodes. See Appendix Table I for a listing of the trade name and manufacturer of fenamiphos.

8

Fairy Rings and Localized Dry Spot

Introduction

Fairy rings is the name commonly used for the circles of mushrooms and puff balls or circular bands of rapidly growing grass that develop in established turf. The term "fairy rings" has its origin in the myths and superstitions associated with their occurrence in the Middle Ages. During these times, the circles were thought by some to be the dancing sites of fairies. In Holland, the dead grass in the center of the ring was supposed to mark the place where the devil churned his butter. In France, some believed that intentional entrance into one of the rings would result in an encounter with large toads with bulging eyes. If a Scottish farmer tilled an area containing fairy rings, then his life would be filled with problems. A more optimistic view of fairy rings was taken in England, where it was considered a good omen to build a house on land supporting them. Fairy rings are now known to be caused by several species of soil and thatch inhabiting fungi.

There are two basic types of fairy rings: (1) **edaphic**[1] — rings that are produced by fungi that colonize primarily the soil, and (2) **lectophilic**[2]—rings that are produced by fungi that colonize primarily leaf litter and thatch. A problem closely related to lectophilic fairy rings is a hydrophobic condition of the thatch and rooting mixture known as **localized dry spot.**

---

[1]Edphaic from *edaphos* (Gr) referring to the soil or earth as a foothold for higher plants.

[2]Lectophilic from *lectus* (L) meaning bed, litter or thatch, and *philos* (Gr) meaning love of, or favorably disposed toward.

Edaphic Fairy Rings

- **Pathogens:**

Fifty-four species of mushrooms (Agaricales) and puff balls (Lycoperdales)

- **Grasses Affected:**

All cultivated warm season and cool season turfgrasses

- **Season of Occurrence:**

Spring, summer and fall

- **Symptoms and Signs:**

In overall view, zones of stimulated grass develop as more or less continuous, circular bands of turfgrass that are darker green and faster growing than the adjacent plants of the same species. These belts of greener plants may range from 4–12 inches (10–30 cm) wide, and the diameter of the circles they form will usually vary from 3–12 feet (0.9–3.7 m). Several distinct rings frequently occur in the same area. In these cases, as the rings converge on each other, fungus activity ceases in the zones of contact. As the result, the concentric shape of the original rings gives way to a scalloped effect.

A characteristic feature of fairy rings is the presence of the fruiting bodies of the associated fungi (sporophores) in the band of stimulated turfgrass. Commonly referred to as "mushrooms," "toadstools," and "puffballs," from time to time these structures may be abundant throughout the circumference of the rings (Plate 8-1; Plate 8-2; Plate 8-3).

- **Conditions Favoring Disease Development:**

Edaphic fairy rings are produced by fungi that colonize primarily the soil. Depending on the nature of the soil profile and the presence or absence of organic matter, these fungi may go to a depth of 2–3 feet (0.6–1 m). Edaphic fairy rings begin from transported bits of mycelium. Two to three years from the beginning of active colonization of the soil are usually required for the characteristic rings to develop.

Although the diameters of typical edaphic fairy rings range from 3 feet (1 m) to 12 feet (3.7 m), rings have been found that measure up to 2600 feet (800 m) across. Variations in rate of enlargement of the rings have been recorded from 3 inches (7.6 cm) to 2 feet (0.6 m) per year. Areas void of vegetation, or currently supporting active fairy rings, serve as barriers to the outward progress of the rings.

- **Control:**

Success in the control of fairy ring with fungicidal applications is marginal at best. The hydrophobic condition of the soil makes it difficult to reach and permeate the mycelial mass with aqueous solutions or suspen-

Plate 8-1. Edaphic fairy ring showing early symptoms with fruiting bodies of the causal fungus.

Plate 8-2. Early symptoms of edaphic fairy ring with fruiting bodies of the causal fungus.

Plate 8-3. Edaphic fairy ring showing advanced symptoms.

sions of fungicides—even when penetration is aided by coring and the use of wetting agents. Also, if control is achieved, it is often short lived.

The most reliable techniques for control of fairy ring are (1) fumigation with methyl bromide, or (2) prolonged soaking of the soil with water.

1. Fumigation with Methyl Bromide—Methyl bromide is an effective fairy ring control fumigant; however, proper use of the material requires a maximum of skill and precaution. It can not be overemphasized that methyl bromide in its final form is a poisonous gas; therefore, when the material is being used, extreme caution should be exercised to keep all children and pets away from the area.

2. Prolonged Water Soaking of Soil—This system consists of water soaking the soil for a distance of 18 inches on either side of the stimulated zone and to a depth of 12 inches (30 cm), and maintaining this condition for a period of 4–6 weeks. A hydrogun or a tree root feeder is ideal for establishing the water soaked condition. If these are not available, and surface irrigation is to be employed, then the area to be treated should be uniformly perforated with a hand aerifier or garden fork to a depth of 4–6 inches (10–15 cm) prior to the first application of water. It is important that the soil remain saturated throughout the entire treatment period; therefore, water should be applied to the affected area every second day.

Lectophilic Fairy Rings and Localized Dry Spot

• **Pathogens:**
Several species of mushrooms (Agaricales)
• **Grasses Affected:**
All cultivated warm season and cool season turfgrasses
• **Season of Occurrence:**
Spring, summer and fall
• **Symptoms and Signs:**
The symptom patterns for lectophilic fairy rings range from the development of either (1) arcs or complete circles of mushrooms with no apparent effect on the plants, (2) circular bands of darker green grass, (3) circular bands of yellow grass, or (4) circular bands of dead turfgrass. Occasionally, white mycelial growth can be observed on the lower leaves of affected plants, permeating the thatch layer to a depth of 0.25 inch (7 mm). The belts of affected grass may range from 2–4 inches (6–10 cm) wide, and the diameter of the circles they form will usually vary from 1–6 feet (0.3–1.8 m). Several rings often develop simultaneously in the same area. In these cases, if they converge, fungus activity ceases at the areas of contact. As the result, the concentric shape of the original rings changes to a scalloped effect. The infested thatch often develops a strong "mushroom" odor. Also, the growth rate of young roots developing at the nodes may temporarily be slowed down (Plate 8-4).

Plate 8-4. Lectophilic fairy ring in early stages of development on creeping bentgrass under golf green management.

In some instances, the thatch and underlying soil within the zones of activity of lectophilic fairy rings becomes hydrophobic. This condition appears to be brought on by the coating of large, individual soil particles by fungus-engendered, hydrophobic material. Although these "localized dry spots" are usually thought of as problems unique to sand-based golf greens, they can also develop on greens that have been constructed with soil mixtures and in fairway and landscape turf (Plate 8-5).

Not all lectophilic fairy rings lead to the development of localized dry spots, nor are these water repellant zones always preceded by visible ring formation. Therefore, although certain lectophilic fairy ring species appear to be involved in the formation of localized dry spots, it is highly possible that this phenomenon can be brought on by a much wider range of thatch-inhabiting microflora.

• **Conditions Favoring Disease Development:**
The primary activity of lectophilic fairy ring fungi takes place in the leaf litter and thatch. Lectophilic fairy rings can occur on turf under any type of management, but are best known for their development on bowling greens and golf course putting greens. Their impact on the turf ranges from little or no damage to the production of unsightly rings to the development of circular patches of dead grass.

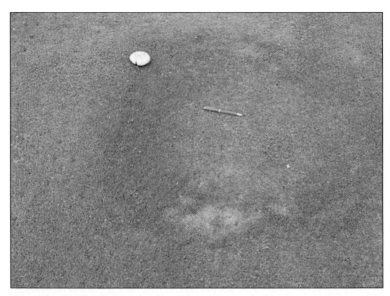

Plate 8-5. Lectophilic fairy ring showing stimulated grass and beginnings of localized dry spot.

Since the species that form lectophilic fairy rings are thatch and litter colonizers, factors that affect the thickness and general condition of the thatch will have a direct bearing on their development. Also, soil sterilization procedures, and/or the use of certain fungicides that lower the populations of either competitors or antagonistic soil and thatch-inhabiting microorganisms on established turf may bring about an increase in the formation of lectophilic rings.

*   **Control:**

1. Cultural Practices—Prevention of the accumulation of heavy thatch layers through the use of vertical cutting and coring equipment will reduce the incidence of lectophilic fairy rings. The hydrophobic condition created by the presence of the fairy ring fungi can be alleviated by a combination of coring and the use of surfactants. Once wetting of the thatch and underlying soil has been accomplished, it is very important that measures be taken to maintain this profile in a continuously moist condition.

2. Chemical Control—Drench applications of flutolanil have been reported to be effective in controlling the development of lectophilic fairy rings and reducing the severity of localized dry spot. See Appendix Table I for a listing of the trade name and manufacturer of flutolanil.

Maximum control from flutolanil requires that it be applied as follows: The day prior to fungicide application, spike or aerify the area to be treated, apply a surfactant, and then irrigate with sufficient water to bring the rooting mixture to field capacity (-0.033 MPa) to a 6-inch (15 cm) depth. Twenty-four hours later apply the fungicide at the label rate for fairy ring control in 1–2 gallons of water per 1,000 square feet (4.2–8.4 liters/93 m^2). Before the leaves dry, irrigate the treated area with 1/4–1/2 inches (10–20 cm) of water. The thatch should then be kept moist for 2–3 weeks.

Notes

9

Diseases Caused by Viruses and Prokaryotes

Introduction

Viruses are submicroscopic, infectious particles that multiply only within living host cells and are pathogenic. The primary pathogenic effects of viruses on plants are not caused by the depletion of the host's nutrients during the synthesis of virus particles. Instead, the metabolic processes of the affected cells are altered to the extent that in addition to synthesizing virus particles, they also produce compounds which interfere with normal cell functions. For example, certain plant viruses can cause the host to produce enzymes that cause necrotic spots and streaks to develop in its leaves.

Viruses can also cause a significant alteration in respiratory rates in plant cells. Some viruses either destroy chloroplasts or interfere with their formation, causing mottling and yellowing of leaves. Others can bring about a change in the plant's production of growth regulators, causing it to either become stunted or slow to recover after clipping. At present, eight virus diseases of turfgrasses have been identified. Of these, St. Augustinegrass/centipedegrass decline is the only one known to cause major damage to the turf.

Prokaryotes are single-celled microorganisms. Instead of a cell wall, they have an envelope surrounding the cytoplasm. Two forms of prokaryotes are known to cause disease in turfgrasses, **bacteria** and **mollicutes.** The cell membrane of the majority of species of plant pathogenic bacteria is made up of a rigid wall with outer and inner membranes. Mollicutes, on the other hand, are much smaller than bacteria and they are covered with a highly pliable membrane. Two important diseases of turfgrasses are known to be incited by prokaryotes, bacterial wilt of cool season grasses and white leaf of bermudagrass.

St. Augustinegrass/ Centipedegrass Decline

Pathogen:

Panicum mosaic virus (a spherical RNA virus)

- **Grasses Affected:**

St. Augustinegrass (*Stenotaphrum secundatum*) and centipedegrass (*Eremochloa ophiuroides*)

- **Season of Occurrence:**

Spring, summer and early fall

- **Symptoms:**

Foliar symptoms develop 7 to 21 days from the time of infection. Rate of symptom development appears to be more rapid at lower air temperatures. The initial symptoms on St. Augustinegrass are a mild chlorotic mottling of the leaf blades which soon progresses to a distinctive stippling effect. With centipedegrass, early symptoms of the disease appear as light green to yellow blotches on the leaves (Plate 9-1). During the second year after infection, a general chlorosis of the leaves develops on both grass species, and there is a retardation in the rate of stolon growth.

Plate 9-1. Close view of individual leaf symptoms of St. Augustinegrass decline.

With the advent of the third year following infection, necrosis of the leaves and stolons and death of plants occurs. The severity of leaf necroses is often extensive enough to bring about the death of areas of turf ranging from 3 to 30 feet (1–9 m) in diameter (Plate 9-2; Plate 9-3).
• **Conditions Favoring Disease Development:**
Rotary mowers are believed to serve as primary means of spread of the virus within established turf. Panicum mosaic virus is introduced into new areas by means of diseased sod. Plants affected by St. Augustinegrass/centipedegrass decline are more vulnerable to herbicide injury and damage from low air temperatures.
• **Control:**
1. Chemical Control—The close linkage between virus activity and the metabolism of the host cell makes it very difficult to develop a chemical that can specifically interfere with virus multiplication without detrimental effects on the normal functions of the host. Therefore, chemical control of St. Augustinegrass/centipedegrass decline is not yet practicable.
2. Management Practices—Diseased St. Augustinegrass and centipedegrass will respond temporarily to fertilization with a complete fertilizer plus iron. The response, however, will not be long lasting and symptoms of the disease will remain.

Plate 9-2. Overview of symptoms of St. Augustinegrass decline.

Plate 9-3. Overall view of symptoms of St. Augustinegrass decline.

When installing a new turf, use only sod that is certified to be free of this disease. Also, care should be taken to reduce the likelihood of transport of virus-bearing plant parts from a diseased stand to healthy turf on mowing, coring and aerification equipment.

The St. Augustinegrass cultivars 'Floratam' and 'Raleigh' are resistant to Panicum mosaic virus. These varieties may be plugged into a diseased stand for a gradual transition to healthy turf. They will fill in as the existing sod thins out. It should be noted, however, that 'Floratam' has very poor cold tolerance and should only be used in the southernmost portions of the United States.

Bacterial Wilt

- **Pathogen:**
Xanthomonas campestris pv. *graminis*
- **Season of Occurrence:**
Spring, summer and early fall
- **Grasses Affected:**
Colonial bentgrass (*Agrostis tenuis*), creeping bentgrass (*Agrostis palustris*), annual bluegrass (*Poa annua*), Kentucky bluegrass (*Poa pratensis*), red fescue (*Festuca rubra*), tall fescue (*Festuca arundinacea*), hard fescue (*Festuca ovina* var. *duriuscula*), annual ryegrass (*Lolium multiflorum*), and perennial ryegrass (*Lolium perenne*)
- **Symptoms:**
In overall view, affected creeping bentgrass golf greens develop an uneven, mottled appearance with irregularly shaped green areas interspersed throughout large sections of reddish-brown turf (Plate 9-4; Plate 9-5; Plate 9-6). The leaves wilt from the tip downward toward the sheath. Within 24 to 48 hours from the onset of wilting, the leaves turn blue green and become shriveled and twisted, after which they become a reddish brown. During the early wilting stage, root and crown tissues are white and healthy appearing. However, soon after the death of the leaves,

Plate 9-4. Overall view of bacterial wilt of C-15 Toronto bentgrass.

Plate 9-5. Overall view of bacterial wilt of Toronto C-15 bentgrass.

brown necrotic streaks develop in their vascular tissue. Ultimately death and decomposition of the entire plant occurs.

On ryegrasses, tall fescue, and Kentucky bluegrass, leaves develop chlorotic and then necrotic stripes along the veins and leaf margins. The striping usually extends from the sheaths upward throughout the entire length of the blades. At high air temperatures, young leaves curl and wither without discoloration or lesions. During warm weather, plants breakdown rather quickly, and the phenomenon is often attributed to summer drought.

The "ooze test" can be used as a aid in diagnosing bacterial wilt. This procedure consists of placing a small piece of stem or leaf on a microscope slide. The tissue should then be cut transversely with a sharp instrument, immersed in a drop of water, and examined immediately with a microscope. If the disease is bacterial wilt, masses of bacteria will be seen streaming from the vascular tissue.

• **Conditions Favoring Disease Development:**
The bacterium overwinters in diseased plants. Primary infections occur through wounds. Cultural practices such as mowing and coring do not appear to be a major factor in the spread of the disease within creeping bentgrass golf greens. Uniform distribution of bacterial wilt on bentgrass

Plate 9-6. Overall view of advanced stage of bacterial wilt of Toronto C-15 bentgrass under golf green management.

greens occurs only when the turf has been established from diseased stolons.

The pathogen colonizes vascular tissue in the leaves, stems, roots and crowns. Large masses of bacterial cells develop in the vessels thus preventing the normal movement of water through the plants. This leads to wilting and necrosis of meristematic tissue at the margins of the leaves and in the leaf ribs.

There is a relationship between the height of cut of creeping bentgrass golf greens and the severity of bacterial wilt. The disease is less severe in the collars than on the putting surface. Also, greens cut at 2/16 inch (2 mm) are more severely affected by the disease than when the putting surface is maintained at higher cuts. The disease is most severe on bentgrass golf greens following periods of cool air temperatures and continuing rainfall, particularly where soils drain slowly. Also, occurrences of bacterial wilt are usually more severe when these periods of rainfall are followed by 2 or 3 days of full sunshine.

• **Control:**

1. Cultural Practices—Golf or bowling greens afflicted with bacterial wilt should be maintained at a cutting height of 3/16 to 1/4 inch (3–4 mm). The only satisfactory long-term solution is replacement of the affected turf with sod or stolons that have been certified to be free of the causal bacterium. When creeping bentgrass is propagated vegetatively, the stolons, or sod that has been produced from stolons, should be checked prior to use for the presence of the bacterial wilt pathogen.

2. Use of Bactericides—Using high rates of the antibiotic oxytetracycline (Mycoshield™, Pfizer Corporation) has been found to bring about a remission of the symptoms of bacterial wilt. However, since these effects are only temporary and the individual treatments are very expensive, use of the material on a field scale for control of the disease is not feasible.

Setaria viridis
(bottle grass, green bottle grass, green bristlegrass,
green foxtail, pigeon grass, wild millet)

White Leaf of Bermudagrass

- **Pathogen:**

Spiroplasma citri

- **Season of Occurrence:**

Spring, summer and early fall

- **Grasses Affected:**

bermudagrass (*Cynodon dactylon*)

- **Symptoms:**

In overall view, white leaf is seen as isolated patches of light yellow to off-white turf measuring 0.5 to 2 inches (1.3–5 cm) in diameter. The internodes of the upper portions of the stems of affected plants are shortened, causing the leaves to be massed at the terminal ends of the stems, creating a "witches' broom" effect. The leaves of affected plants are at first pale yellow. They then turn a distinctive dull white. Also, the leaves of diseased plants are slightly broader and flatter than normal (Plate 9-7; Plate 9-8).

- **Disease Profile:**

Apparently the disease does not cause a severe thinning or loss of turf. Its primary impact is aesthetic. The pathogen is thought to be transmitted by leaf hoppers.

Plate 9-7. White leaf of bermudagrass under golf green management.

Plate 9-8. White leaf of bermudagrass under golf green management.

- **Control:**

At present, there are no reports of successful control of white leaf with a pesticide. The only satisfactory long-term solution is when establishing new bermudagrass turf, use stolons or sod that is known to be free of the spiroplasma that causes this disease.

Notes

10

Surface Algae

Introduction

Of the various surface-inhabiting algae associated with turf, the blue-green (Cyanobacteria) forms are generally considered the most troublesome. The shape of blue-green algae varies from unicellular rods or spheres to filamentous strands in which the individual cells are arranged end-to-end (Plate 10-1). Their color is derived from varying amounts of chlorophyll *a* (green pigment), carotenoids (yellow pigment), phycocyanin (blue pigment), or sometimes phycoerythrin (red pigment) contained in the cells. Consequently, not all so-called "blue-green" algae are blue-green—black, brown, yellow, red, and grass-green colors may also occur.

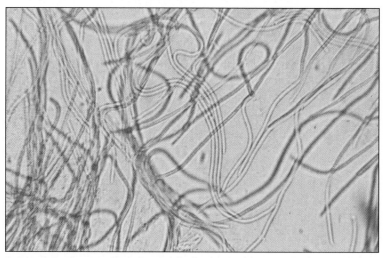

Plate 10-1. Individual cells of the blue-green algae, *Oscillatoria tenuis.*

Season of Occurrence
Algae are often a major problem in turfgrass management during warm
weather when there are frequent rainfalls and the stand of grass begins to
thin due to environmental stress and root and foliar disease problems.

Symptoms and Signs
The occurrence of blue-green algae is marked initially by dark scum or
slime on the surface of the soil. The color may vary from black, to green,
to dark green, to green-brown depending on the species involved. When
wet, the scum has a sticky or rubbery texture. When dry, it may form a
parchment-like crust which in turn may peel and loosen from the soil.
Also, turf in which the soil surface is heavily colonized by algae may become yellow. This is thought to be due to either iron chlorosis or the production of root toxins (Plate 10-2).

Effects of Algae on Turfgrass Growth and Development
The species of blue-green algae found in conjunction with turf do not parasitize the plants; their effect on grass growth and development is indi-
rect. They produce a mucilaginous slime which covers the soil surface,
repelling water and chemical solutions. This slime also moves into the
soil's pores, plugging them and thus impeding water infiltration and re-

Plate 10-2. Bluegreen algae colonizing the thatch and soil surface of
creeping bentgrass under golf tee management.

ducing aeration. Algal scum will in itself reduce turfgrass vigor and make the establishment of seedlings difficult. Finally, when the slime layer dries, it forms a crust which can further hamper the reestablishment of grass plants in the affected area.

Conditions Favoring Algal Growth
Surface water—A major factor in algal colonization of turf is water status. High populations of algae commonly occur when the soil surface stays wet for long times due to either poor surface drainage, low infiltration rates, or frequent rains. The colonization of soil surfaces by algae is common on the lower sections of sloping ground. This is probably due to extended periods of wetness brought on by additional water flowing across the soil surface.

    Light—The growth of blue-green algae is favored by high levels of sunlight. Although abundant algal growth is commonly found in shaded areas, the primary effect of shading is reduction in evaporation rates, allowing the soil surface to remain wet for a longer period of time. Algal growth is more prolific on golf greens in which are managed under ultra low mowing heights in order to meet requirements for faster and more uniform putting surfaces—a practice that allows more light to reach the soil surface.

    Soil compaction—Algal growth is greater in turf growing on compacted soils. This is due to (a) the reduction in overall grass growth thus allowing more light to reach the soil surface, and (b) a prolonging of the duration of soil surface wetness by reducing the rate of water infiltration into the soil.

    Soil pH—Acidic soils favor growth and development of algae.

    Soil fertility—Algal growth is usually more prolific when the turf is grown under low fertilization. This is a consequence of poor grass growth, thus exposing the soil surface to more light. Where types of fertilizer is concerned, it has been observed that the use of natural organic fertilizers tends to promote the growth of algae. Also, frequent fertilization with ammonium sulfate lowers the soil pH which in turn fosters an increase in algae.

Control:
1. Cultural Practices—An important aspect of algae control is the rapid removal of water from the soil surface by (a) improving overall surface drainage and (b) increasing penetration and infiltration rates by coring and vertical mowing to maintain thatch levels at 1/2 inch (13 mm) thickness.

2. Preventive Procedure—Apply chlorothalonil or mancozeb sprays on a 7–14 day schedule at the low label rates listed for algae control. If the conditions are particularly favorable for the growth of algae, the fungicide should be applied at the high label rate and on a 7-day schedule. Chemical treatments are more effective when applied to relatively dry turf.

3. Eradicative Procedure—When algae is well established, an attempt should be made to dry out the soil surface of the affected area. The area should then be raked, spiked or verticut to break up the algal crust and facilitate grass recovery. This should then be followed by a light top dressing with an appropriate top dressing mixture or a wicking agent such as Profile^TM1 to help maintain a dry soil surface. Then the area should be treated with either chlorothalonil or mancozeb at the curative label rates in 3 to 5 gallons (11–19 liters) of water per 1000 square feet (93 m^2). Two or three rounds of spray at 7-day intervals are usually required to establish satisfactory control.

For a listing of representative trade names and manufacturers of chlorothalonil and mancozeb, see Appendix Table I.

---

[1]A product of AIMCOR 750 Lake Cook Road Suite 440 Buffalo Grove, Illinois 60089

11

How to Develop Integrated Turfgrass Disease Control Strategies

Introduction

Maximum effectiveness in turfgrass disease control is achieved through strategies that incorporate all factors known to reduce the severity of the target diseases. In developing these programs, the turfgrass management specialist must not only carefully consider the effect each component will have on the aesthetic and utilitarian standards that have been defined for the turf, but also evaluate the impact it will have on both environmental quality and public health.

The basic resources for diminishing the severity of turfgrass diseases fall into one of four categories: (1) disease resistant grasses, (2) cultural practices, (3) biological control agents, and (4) chemical pesticides

Use of Disease Resistant Grasses

The various species of grass used for sports, recreation and landscape turf show a wide range of variability in their susceptibility to specific diseases. For example, (a) annual bluegrass and Kentucky bluegrass are highly susceptible to **melting-out** but tall fescue and perennial ryegrass are very resistant, (b) annual bluegrass and Kentucky bluegrass are very susceptible to **summer patch,** fine fescues are moderately susceptible, and creeping bentgrass and perennial ryegrass are highly resistant to the disease, and (c) with **Fusarium blight,** annual bluegrass and bentgrasses show the highest degree of susceptibility to the disease, Kentucky bluegrass ranks next in order of susceptibility, and perennial ryegrass is the most resistant species.

A major objective of turfgrass breeding programs is developing disease resistant varieties. Lists showing the comparative levels of disease resistance of commercially available varieties may be obtained from turfgrass seed producers and local turfgrass management advisory services.

The degree of resistance of the species and/or variety under cultivation to commonly occurring diseases in the area in question will determine how strong the fungicide component of the control program needs to be. In temperate climates, annual bluegrass dependent turf will require a more intensive fungicide program for control of both growing season and winter diseases than if the predominant species under management were either Kentucky bluegrass or creeping bentgrass. Also, a stronger fungicide program is required to effectively control Rhizoctonia blight and Pythium blight on creeping bentgrass and perennial ryegrass in hot and humid subtropical climates than when they are grown in relatively cool temperate zones.

Cultural Practices

Nutrition and Soil pH

Of the various nutrient elements, nitrogen has the greatest impact on turfgrass disease development. For example, Helminthosporium leaf and crown diseases are more severe under high nitrogen fertilization. In the installation of sod, the use of high rates of nitrogen-based fertilizer to facilitate rooting can bring about a significant increase in the severity of Rhizoctonia blight, Pythium blight, and Fusarium blight, and late fall fertilization with a readily available source of nitrogen will increase the severity of the winter diseases Fusarium patch of cool season grasses and spring dead spot of bermudagrass.

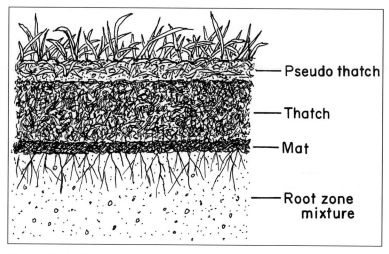

Figure 11-1. Cross section of sod showing the layers that develop between the soil surface and the zone of green vegetation. The pseudothatch is formed from partially decomposed leaf clippings; the thatch is made up of living and dead grass stems, leaves, and roots; and mat is a mixture of soil, stems, and roots. *From Couch, 1995.*

Soil pH can also have a significant effect on the severity of certain turfgrass diseases. The severity of summer patch, take-all patch, Fusarium patch and spring dead spot are greater under alkaline soil conditions.

Thatch Management
Thatch is the layer of living and dead grass stems, leaves and roots that develops between the soil surface and the zone of green vegetation. When this layer becomes intermixed with soil, it is referred to as a **mat**. A thin mat layer often forms in the section of thatch immediately adjacent to the soil surface. However, extensive applications of top dressing, or when the turf becomes covered with silt as the result of flooding, often transforms the entire thatch into a mat. A less densely packed layer of partially decomposed leafclippings known as the **pseudothatch** frequently forms on the surface of the thatch or mat body (Figure 11-1).

Thatch serves an important role in protecting turfgrass crowns and leaves from the shearing action of foot traffic and being compressed by the wheels of maintenance equipment and golf carts. Generally speaking, 1/2 inch (1.3 cm) is considered an appropriate thatch thickness where protection of the turf from wear is concerned.

Amounts of thatch greater than 3/4 inch (1.9 cm) in thickness can result in turf that is more susceptible to stresses from the physical environment and the development of fungus-incited diseases. Under conditions of heavy thatch, the vital crowns, lateral stems, and roots form in the thatch layer, rather than developing in the soil. When this happens, the plants are more vulnerable to heat, cold, and drought injury.

Many of the fungi that parasitize the leaves and crowns of turfgrasses also grow as saprophytes in the thatch, pseudo thatch and mat zones. Thatch buildup means increased amounts of inoculum and thus increased disease incidence and severity. Fusarium patch, spring dead spot of bermudagrass, Fusarium blight, Helminthosporium leaf spot, Sclerotium blight, Sclerotinia dollar spot, and Pythium blight are a few examples of major diseases of turfgrasses that are more severe under heavy thatch conditions.

Mowing Height
Generally speaking, there is an inverse relationship between mowing height and the severity of turfgrass diseases. As the height of cut of sports, recreation, and landscape turf is lowered, its susceptibility to disease increases. For example, the severity of spring dead spot of bermudagrass is greater in turf maintained at low cutting heights; Helminthosporium leaf spot of Kentucky bluegrass is more severe on plants cut at less than 1 1/2 inches (3.8 cm); and the incidence of take-all patch is higher on creeping red fescue cut at 3/4 inch (1.9 cm) than when mowed at 1 1/2 inches (3.8 cm).

Soil Moisture Stress
The reference points commonly used in descriptions of soil water are saturation capacity, field capacity and permanent wilting percentage. **Saturation capacity** is the moisture content of a soil when all pores are filled with water. **Field capacity** is the moisture content of a soil after gravitational water has drained away. It is the amount of water retained only by the soil's capillary pores. It is the upper limit of total water available for plant growth over an extended period of time. In a well drained soil, field capacity is usually reached 2 to 5 days after the profile is wetted by rain or irrigation. **Permanent wilting percentage** is reached through plant extraction. It is the moisture content of the soil at the time the leaves wilt and will not recover their turgidity unless water is added to the soil. As such, then, it is the lower limit of soil water available for plant growth.

Soil moisture contents from field capacity to permanent wilting percentage have a significant effect on the growth and development of turfgrass. Beginning at field capacity, as the moisture content of the soil adjacent to fibrous roots is extracted by the plant, transpiration and photosynthesis rates are lowered, and metabolic actions are set in motion that bring about an alteration in the relationship between starches and sugars and between amino acids and proteins. When the permanent wilting percentage is reached, cell turgor is lost and stomata close.

These various changes can in turn affect the susceptibility of turfgrass to infection and colonization by pathogenic microorganisms. Outbreaks of Fusarium blight, Pythium blight, and Sclerotinia dollar spot are more severe when the turf is growing at soil moisture levels at less than 3/4 field capacity.

Duration of Leaf Wetness

The presence of free water on the surface of the leaves is an important factor in the development of foliar diseases of turfgrasses. Through guttation fluids and leachates from underlying tissue, leaf surfaces are subject to deposits of a wide range of chemical compounds including vitamins, proteins, amino acids, and sugars. When these materials become dissolved in the leaf surface water, they serve as a source of nutrition for invading fungi. Moisture that has accumulated on the leaves as the result of light, intermittent rainfalls or condensation of dew or fog will contain a higher concentration of these compounds than free water that has developed from heavy, extended rainfalls.

The length of time leaves are wet will directly affect the number of infections per leaf. When air temperatures are optimum for growth of a parasitic fungus, a much shorter duration of leaf wetness is required for infection to occur than at temperatures outside this range. In general, when the level of inoculum is high and the day-night temperatures are optimum for growth of the pathogen in question, one wet leaf period of 12 to 24 hours duration will usually result in a significant outbreak of disease. On the other hand, if the initial inoculum level is low, then a close succession of two to three wetting periods of 48 to 72 hours each is usually necessary before the disease reaches serious proportions.

When the climatic conditions are such that guttation fluids and the condensation of fog and dew are deposited on the leaves throughout the night, practices should be followed that reduce the durations of leaf wetness to less than 12 hours. This can be accomplished by the early morning removal of the leaf surface water by using a large brush-type broom or by dragging a water hose across the area. Also, the duration of leaf wetness due to these conditions can be shortened 2 to 4 hours by following

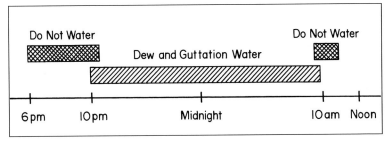

Figure 11-2. Illustration of how watering schedules in temperate climates impact on the duration of daily periods of leaf wetness. *From Couch, 1995.*

a nighttime watering schedule in which the irrigation system is programmed to begin at least 3 hours after sunset and be completed before sunrise (Figure 11-2).

Biological Control
The basic concept of biological control of plant diseases is to reduce the incidence and severity of the target disease by the use of a microorganism that either decreases the population density of the pathogen or significantly reduces its disease producing activities. Where turfgrass diseases are concerned, attempts at biological control fall into one of two categories: (1) use of "biopesticides," *i.e.,* preparations containing known microbial species that are detrimental to the growth and development of specific pathogens, and (2) use of natural organic materials colonized by complex mixtures of unidentified microbial species that may have the potential to suppress the development of pathogenic fungi and nematodes.

University-based tests have shown unsatisfactory levels of success in disease control when turf is treated with either experimental or presently available commercial biopesticides. However, with improved expertise in establishing and maintaining high populations of microorganisms in the thatch and soil that are selectively antagonistic to the pathogenic species, this approach to disease control may eventually become feasible.

Also, research evidence to date has shown that natural organic materials such as kelp, granular preparations of sea plants, composts, sewage sludge, organic fertilizers, and manure-based products do not significantly reduce the severity of turfgrass diseases. In addition, when compared simultaneously at the same test sites, natural organic, synthetic organic, and mineral fertilizers have all been found to have (a) identical impacts on the incidence and severity of specific diseases, and (b) similar effects on microbial development in the thatch and soil.

Use of Chemical Pesticides

Chemical pesticides are an essential component of successful turfgrass disease control strategies. Although the number of products available for control of root parasitic nematodes in established turf is limited, a wide variety of chemicals is available for the management of fungus-incited diseases.

Pesticide Nomenclature

Three names are used in commerce to designate a given pesticide: (1) a chemical name, (2) a common name, and (3) proprietary or trade name. The **chemical name** is usually very complex and while it helps standardize communication among scientists, it often has little meaning to the user of the product. The **common name** is assigned to the compound to avoid the difficulties the consumer often has in remembering or pronouncing the chemical name. Common names cannot be copyrighted. They are developed by designated national committees and are recognized on an international basis as the standard name for the pesticide and can only be used to designate that particular chemical compound. The **trade name** is a product name that has been coined and copyrighted by either the manufacturer or the marketing agency of a particular pesticide. It is not unusual for a given pesticide to have more than one trade name when it is being sold (a) by different marketing agencies, (b) for use on different commodities, or (c) for use on the same commodity but in different nations or regions of the world. The following is an example of the full nomenclature of a specific turfgrass fungicide:

Chemical Name: 3-[3,5-dichlorophenyl]-N-[1-methyl]t,4-dioxo-
                1-imidazolidinecarboxamide
Common Name: Iprodione
Trade Name: Chipco 26019™

A widely used system of classifying fungicides is based on their modes of action, biochemical and topical. **Biochemical mode of action** refers to the effects of the fungicide on the metabolic processes of the fungal cell. **Topical mode of action** identifies the location in or on the plant in which the fungitoxic or fungistatic activities take place. The topical modes of action of fungicides fall into one of two categories, contact and penetrant.

Contact fungicides are active only at the site of placement. For example, a contact fungicide sprayed onto the surface of a leaf will only be active against fungi in that immediate location. Chlorothalonil (Daconil 2787™) and mancozeb (Fore™) are examples of contact fungicides.

Penetrant fungicides are active at the site of placement but they also penetrate into the underlying plant tissue in amounts that are toxic to the invading fungus. Three forms of penetrant fungicides are used in turfgrass disease control: (1) localized penetrants, (2) acropetal penetrants, and (3) systemic penetrants.

- **Localized penetrants** pass into the underlying tissue in fungicidal amounts but they remain in the area if of initial entry. Iprodione (Chipco 26019™) and vinclozolin (Curalan™) are examples of localized penetrants.
- **Acropetal penetrants** are fungicides that pass into the underlying tissue and then are translocated in fungicidal amounts in the xylem (water conducting tissue), thus they are only distributed upward from the site of entry. Mefenoxam (Subdue MAX™), flutolanil (Prostar™) and azoxystrobin (Heritage™) are examples of acropetal penetrants.
- **Systemic penetrants** pass into the underlying tissue and then are translocated simultaneously in fungicidal amounts upward in the xylem and downward in the phloem (plant metabolite conducting tissue), thus they are distributed uniformly throughout the plant. Fosetyl Al (Aliette™) is the only systemic penetrant fungicide presently marketed for turfgrass disease control.

Contact fungicides usually have a broad spectrum of disease control. They are less vulnerable to resistance than penetrants but usually provide control for shorter durations of time. One of the reasons penetrant fungicides are often longer lasting than contacts is that they are not subject to being dislodged by rainfall or irrigation.

The common name, topical mode of action, and a representative trade name and manufacturer for each currently marketed turfgrass fungicide are listed in Appendix Table I.

Minimizing the Risk of Fungicide Resistance
The development of individual cases of fungicide resistance can result in extensive damage to a stand of turfgrass before alternative fungicides can be placed into use. This is the reason that turfgrass disease control programs should always include tactics for minimizing this risk.

The most effective strategy for reducing the likelihood of fungicide resistance is one that rotates among fungicides with different biochemical modes of action and utilizes all available resources to provide maximum disease control. Specifically:

1. The same fungicide should not be used for more than three successive applications before switching to a fungicide with a different mode of action.
2. Fungicides should not be used as single component sprays at less than the manufacturers' low label rates.
3. When possible, use fungicide mixtures that have been shown to be synergistic in the control of the target disease.
4. Tank mixtures at less than the low label dosage levels unless the combinations have been shown to be synergistic in the control of the target disease.
5. When possible, apply fungicides on a preventive control schedule rather than relying on curative treatments.
6. The spray should be applied uniformly over the area being treated. Edges of the spray swaths that consistently receive less fungicide are ideal locations for the buildup of resistant populations of pathogenic fungi.

Optimum Fungicide Dilution Levels

Dilution levels can significantly affect the performance of fungicides. Some fungicides are highly dilution specific, while others are effective over a fairly wide range of dilutions (Table 11.1). When the optimum dilution rate is not listed for the fungicide in question, then the application should be made in 1 to 2 gallons of water per 1,000 square feet (4.2–8.4 liters/93 m^2).

Table 11.1. Optimum dilution levels of certain fungicides used at label dosage rates for the control of turfgrass diseases. *From Couch, 1995.*

Fungicide	Optimum Dilution per 1000 ft² (93 m²)
Chlorothalonil (Daconil 2787™)	1.0 gal (4.2 l)
Triadimefon (Bayleton™)	2.0 gal (8.4 l)
Iprodione (Chipco 26019™)	0.5–4.0 gal (2.1–8.4 l)
Propiconazole (Banner™)	2.0 gal (8.4 l)
Vinclozolin (Vorlan™, Curalan™)	1.0–2.0 gal (4.2–8.4 l)

At times, control recommendations may either call for a drench application of a fungicide or state that it should be watered into the turf after application. The purpose for using a large amount of water is to make certain that the treatment reaches the crown and upper 1 inch (2.54 cm) level of the root zone. A drench is a dilution level of 20–30 gallons (76–113 liters) of water per 1000 square feet (93 m²) of turf. Handling this amount of water is feasible where treating small areas is concerned. However, when it becomes necessary to treat a large section, the most workable procedure is to water the fungicide into the turf according to the following procedure.

1. Spike or aerify the area to be treated.
2. Irrigate with sufficient water to bring the rooting mixture to field capacity (-0.033 MPa) to a 6 inch (15 cm) depth.
3. Twenty-four hours later, apply the fungicide at the label rate in 1 to 2 gallons of water per 1000 square feet (4.2–8.4 liters/93 m²).
4. Before the leaves dry, irrigate the treated area with 1/4 to 1/2 inches (10–20 cm) of water.

How to Apply Fungicides Accurately and Uniformly to the Turf Surface
Satisfactory disease control can only be achieved when the fungicides are applied accurately and uniformly to the turf surface. In considering methods of application, it is important to bear in mind that uniform spray patterns on areas of turf cannot be achieved with spray guns. Uniform distribution of spray fungicides over the surface turf can only be accomplished with broadcast booms. The use of spray guns in the application of turfgrass fungicides can be justified; however, when it is necessary to treat either small or isolated locations, or areas in which it is difficult to use broadcast booms.

Nozzle type, pressure at the nozzle, and configuration of the broadcast boom are major factors in establishing its uniformity of application. The best levels of turfgrass disease control are obtained when the fungicides are applied with booms equipped with nozzles fitted with either Delavan[1] LF 80–2, LF 80–4 or Spraying Systems[2] TeeJet™

---

[1]Delavan Inc., 20 Delavan Drive, Lexington, Tennessee 38351 U.S.A.

[2]Spraying Systems Co., North Avenue at Schmale Road, Wheaton, Illinois 60189-7900 U.S.A.

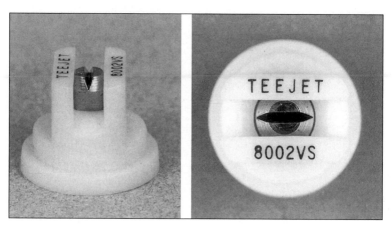

Figure 11-3. Flat fan T-8002 nozzle tip.

8002, 8004 flat fan tips, or with either Delavan RA-10 or RA-15 whirl chamber (Raindrop™) nozzles (Figure 11-3; Figure 11-4). Nozzles fitted with flat fan tips in this size range perform best when fungicides are applied at 30–60 psi (207–414 kPa). Maximum disease control is obtained with Raindrop™ RA-10 or RA-15 whirl chamber nozzles when the fungicides are applied at 40–60 psi (276–414 kPa). The effectiveness of fungicides applied with RA-10 or RA-15 whirl chamber nozzles decreases significantly when the pressure level at the nozzle drops below 30 psi (207 kPa).

Where configuration of the boom is concerned, nozzles fitted with variable spray (VS) flat fan tips should be spaced on the boom to provide a 30 to 50 percent overlap of the spray pattern. If even spray (EVS) flat fan tips are used, some overlap is still necessary.

With whirl chamber nozzles, the boom should be tilted so the nozzles are at a 45 degree angle with the turf surface. Also, it is very important that spacing of whirl chamber nozzles on the boom provide 100 percent overlap of the spray pattern. In each of these instances, the manufacturer's specifications should be checked for proper spacing on the boom and proper boom height.

The amount of spray being delivered per unit area of turfgrass by a broadcast boom is the product of nozzle pressure, nozzle size and speed of the sprayer. Specific procedures for calibrating a power sprayer to deliver the preferred amount of fungicide per 1,000 square feet (93 m²) of turfgrass are outlined in the catalogs of the major manufacturers of spray

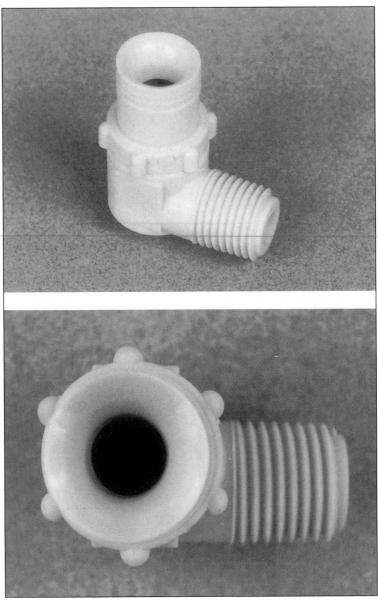

Figure 11-4. Whirl chamber (Raindrop™) RA-15 nozzle.

nozzles. Basically, these procedures consist of measuring the output of one or more of the nozzles within the length of time required for the sprayer to move a predetermined distance, and then subjecting this information to a mathematical formula.

Avoiding skips or major overlaps in the spraying pattern can be accomplished by the use of a boom-mounted unit that deposits small amounts of white foam at spaced intervals. Another approach to assuring the uniformity of applications is the addition of a commercially available polymeric, nonstaining colorant to the spray preparation. This spray indicator helps the operator to maintain the proper amount of overlap between swaths. It also useful in detecting plugged nozzles.

How to Maximize the Effectiveness of Nematicides
For maximum effectiveness, nematicides should not be applied to established turf until the soil temperature at 4 inches (10 cm) deep reaches 60 °F (16 °C) or greater. Perform cultivation practices that improve water infiltration into the soil such as spiking, vertical mowing, or coring. Twenty-four hours prior to the application of the nematicide, irrigate with sufficient water to bring the rooting mixture to field capacity (−0.033 MPa) to a 6 inch (15 cm) depth. Immediately after application of the nematicide, the treated area should be irrigated with 1/2 inch (1.3 cm) of water. Recovery of the turf as a whole is dependent on the production of new roots; therefore, it is important that after the nematicidal treatment has been completed close attention be given to cultivation and watering practices that maximize root growth and development.

Notes

Appendix

Table I. Fungicides and Nematicides Used in the Control of Turfgrass Diseases

The following listing of fungicides and nematicides is arranged alphabetically according to the common (coined) name of each. The trade name included for each has been selected only for the purpose of illustration. No discrimination is intended when the same chemical is marketed under other trade names, nor is endorsement by the author of the products bearing these names implied.

Common (Coined) Name	Topical Mode of Action	Trade Name	Manufacturer's Address
Azoxystrobin	Acropetal penetrant	Heritage™	Zeneca Professional Products 1800 Concord Pike Wilmington, Delaware 19897 Telephone: (800) 759-2500
Chloroneb	Contact	Teremec™	PBI/Gordon Corporation P.O. Box 480 Kansas City, Missouri 64101 Telephone: (816) 421-4070
Chlorothalonil	Contact	Daconil 2787™	Zeneca Professional Products 1800 Concord Pike Wilmington, Delaware 19897 Telephone: (800) 759-2500
Fenamiphos	Systemic penetrant	Nemacur™	Bayer Corporation P.O. Box 4913 Kansas City, Missouri 64120 Telephone: (800) 842-8020
Fenarimol	Acropetal penetrant	Rubigan™	Dow Agrosciences 9330 Zionsville Rd. Indianapolis, Indiana 46268 Telephone: (800) 258-3033
Flutolanil	Acropetal penetrant	Prostar™	Aventis Environmental Sci. Chipco Products 95 Chestnut Ridge Rd. Montvale, New Jersey 07645 Telephone: (800) 843-1702

Table I. Fungicides and Nematicides Used in the Control of Turfgrass Diseases (*Continued*)

Common (Coined) Name	Topical Mode of Action	Trade Name	Manufacturer's Address
Fosetyl Al	Systemic penetrant	Aliette™	Aventis Environmental Sci. Chipco Products 95 Chestnut Ridge Rd. Montvale, New Jersey 07645 Telephone: (800) 843-1702
Iprodione	Localized penetrant	Chipco 26019™	Aventis Environmental Sci. Chipco Products 95 Chestnut Ridge Rd. Montvale, New Jersey 07645 Telephone: (800) 843-1702
Mancozeb	Contact	Fore™	Rohm and Haas Co. 100 Independence Mall West Philadelphia, Pennsylvania 19105 Telephone: (215) 592-3624
Mefenoxam	Acropetal penetrant	Subdue MAXX™	Novartis Crop Protection P.O. Box 18300 Greensboro, North Carolina 27419 Telephone: (800) 334-9481
Myclobutanil	Acropetal penetrant	Eagle™	Rohm and Haas Company 100 Independence Mall West Philadelphia, Pennsylvania 19105 Telephone: (215) 592-3624
Propamocarb	Acropetal penetrant	Banol™	Aventis Environmental Sci. Chipco Products 95 Chestnut Ridge Rd. Montvale, New Jersey 07645 Telephone: (800) 843-1702

Table I. Fungicides and Nematicides Used in the Control of Turfgrass Diseases (*Continued*)

Common (Coined) Name	Topical Mode of Action	Trade Name	Manufacturer's Address
Propiconazole	Acropetal penetrant	Banner MAXX™	Novartis Crop Protection P.O. Box 18300 Greensboro, North Carolina 27419 Telephone: (800) 334-9481
Quintozene (PCNB)	Contact	Terraclor™ Turfcide™	Uniroyal Chemical Co. Benson Road Middlebury, Connecticut 06749 Telephone: (203) 573-3888, *and* 25 Erb Street Elmira, Ontario Canada N3B 3A3 Telephone: (800) 265-2157
Thiophanate Methyl	Acropetal penetrant	Cleary 3336™	W. A. Cleary Chemical Co. 1049 Somerset Drive Somerset, New Jersey 08875 Telephone: (908) 247-8000
Triadimefon	Acropetal penetrant	Bayleton™	Bayer Corporation P.O. Box 4913 Kansas City, Missouri 64120 Telephone: (800) 842-8020
Trifloxystrobin	Acropetal penetrant	Compass™	Novartis Crop Protection P.O. Box 18300 Greensboro, North Carolina 27419 Telephone: (800) 334-9481
Vinclozolin	Localized penetrant	Curalan™	BASF Corporation P.O. Box 13528 Research Tiangle Park, NC 27709 Telephone: (800) 424-9300

Table II. Sources of Local Information on Diagnosis and Control of Turfgrass Diseases in the United States and Canada

United States

ALABAMA

Plant Disease Clinic
Department of Plant Pathology
102 Extension Hall
Auburn University
Auburn, Alabama 36849-5624
Telephone: (334) 844-5508

ALASKA

Department of Plant Pathology
University of Alaska
Agriculture-Forestry Experiment Station
Fairbanks, Alaska 99775-0080
Telephone: (907) 474-7439

ARIZONA

Department of Plant Pathology
Forbes Building #36
University of Arizona
Tucson, Arizona 85721
Telephone: (602) 621-1828

ARKANSAS

Plant Disease Clinic
Lonoke Agricultural Center
P.O. Drawer D; Highway 70 East
Lonoke, Arkansas 72086
Telephone: (501) 676-3124

CALIFORNIA

Department of Plant Pathology
University of California, Davis
Davis, California 95616-8680
Telephone: (916) 752-4269

Table II. Sources of Local Information on Diagnosis
and Control of Turfgrass Diseases in the United States
and Canada (*Continued*)

Department of Plant Pathology
University of California, Riverside
Riverside, California 92521
Telephone: (909) 787-4115

Dr. Larry Stowell
PACE-PTRI
1267 Diamond Street
San Diego, California 92109
Telephone: (619) 272-9897
Web page: www.pace-ptri.com

COLORADO

Plant Diagnostic Clinic
Jefferson County Extension
15200 West 6th Avenue
Golden, Colorado 80401

CONNECTICUT

Consumer Horticultural Center
University of Connecticut
Storrs, Connecticut 06269-4087
Telephone: (860) 486-3437

DELAWARE

Extension Plant Pathologist
Department of Plant & Soil Sciences
136 Townsend Hall
University of Delaware
Newark, Delaware 19717-1303
Telephone: (302) 831-2532

FLORIDA

Nematode Assay Laboratory
Department of Entomology and Nematology
Building 78 Mowry Road
University of Florida
Gainesville, Florida 32611
Telephone: (904) 392-1994

Table II. Sources of Local Information on Diagnosis and Control of Turfgrass Diseases in the United States and Canada (*Continued*)

Plant Disease Clinic
Department of Plant Pathology
Building 78 Mowry Road
University of Florida
Gainesville, Florida 32611
Telephone: (904) 392-1795

University of Florida
IFAS—3205 College Avenue
Ft. Lauderdale, Florida 33314-7700
Telephone: (954) 475-8990

University of Florida
IFAS—EREC
P.O. Box 8003
Belle Glade, Florida 33430
Telephone: (561) 996-3062

GEORGIA

Plant Disease Clinic
4-Towers Building
University of Georgia
Athens, Georgia 30602
Telephone: (706) 542-4102

Department of Plant Pathology
Georgia Station
University of Georgia
Griffin, GA 30223-1797
Telephone: 404/412-4012

HAWAII

Agricultural Diagnostic Service Center
University of Hawaii
1910 East-West Road
Sherman Hall 112
Honolulu, HI 96822
Telephone: (808) 956-2838

Table II. Sources of Local Information on Diagnosis
and Control of Turfgrass Diseases in the United States
and Canada (*Continued*)

IDAHO

Extension Plant Pathologist
Research and Extension Center
University of Idaho
3793 N, 3600 E.
Kimberly, Idaho 83341
Telephone: (208) 423-6603

ILLINOIS

(*May through September*)
Plant Clinic
1401 West St. Mary's Road
University of Illinois
Urbana, Illinois 61801
Telephone: (217) 333-0519

(*October through March*)
Plant Clinic
Department of Plant Pathology
N-533 Turner Hall
1102 S. Goodwin Ave.
University of Illinois
Urbana, Illinois 61801
Telephone: (217) 333-2478

INDIANA

Plant and Pest Diagnostic Laboratory
Department of Botany and Plant Pathology
1155 Lilly Hall of Life Sciences
Purdue University
West Lafayette, Indiana 47907-1155
Telephone: (317) 494-4615

Table II. Sources of Local Information on Diagnosis and Control of Turfgrass Diseases in the United States and Canada (*Continued*)

IOWA

Plant Disease Clinic
Department of Plant Pathology
351 Bessey Hall
Iowa State University
Ames, Iowa 50011
Telephone: (5l5) 294-0581

KANSAS

Plant Disease Diagnostic Laboratory
Department of Plant Pathology
Throckmorton Hall
Kansas State University
Manhattan, Kansas 66506-5502
Telephone: (913) 532-6176

KENTUCKY

Plant Disease Diagnostic Laboratory
Department of Plant Pathology
College of Agriculture
S-305 Agriculture Science Building–North
University of Kentucky
Lexington, Kentucky 40546-0091
Telephone: (606) 257-3901

LOUISIANA

Plant Disease Diagnostic Clinic
220 H. D. Wilson Building
Louisiana State University
Baton Rouge, Louisiana 70803-1900
Telephone: (504) 388-2478

MAINE

Pest Management Office
491 College Avenue
University of Maine
Orono, Maine 04473-1295
Telephone: (207) 581-3880

Table II. Sources of Local Information on Diagnosis
and Control of Turfgrass Diseases in the United States
and Canada (*Continued*)

MARYLAND

Plant Diagnostic Laboratory
Department of Plant Biology
The University of Maryland
College Park, Maryland 20742
Telephone: (301) 314-1611

MASSACHUSETTS

Department of Microbiology
University of Massachusetts
209 E Fernald Hall
Amherst, MA 01003-2420
Telephone: (413) 545-3413

MICHIGAN

Plant Diagnostic Clinic
Department of Botany & Plant Pathology
Room 166 Plant Biology Building
Michigan State University
East Lansing, Michigan 48824-1312
Telephone: (517) 355-4680

MINNESOTA

For homeowners:
145 Alderman Hall
1970 Folweel Avenue
University of Minnesota
St. Paul, Minnesota 55108
Telephone: (612) 825-4642

For professionally managed turf:
Plant Disease Clinic
Department of Plant Pathology
495 Borlaug Hall
1991 Upper Buford Circle
University of Minnesota
St. Paul, Minnesota 55108
Telephone: (612) 625-1275

Table II. Sources of Local Information on Diagnosis and Control of Turfgrass Diseases in the United States and Canada (*Continued*)

MISSISSIPPI

Plant Pathology Laboratory
Room 9, Bost Extension Center
Box 9655
Mississippi State University
Mississippi State, Mississippi 39762
Telephone: (601) 325-2146

MISSOURI

Plant Science Unit/Diagnostic Clinics:
Plant Disease Identification
Room 45 Agriculture Building
University of Missouri
Columbia, Missouri 65211
Telephone: (314) 882-3019

MONTANA

Plant Disease Clinic
Department of Plant Pathology
516 Leon Johnson Hall
Montana State University
Bozeman, Montana 59717
Telephone: (406) 994-5157

NEBRASKA

Plant and Pest Diagnostic Clinic
Department of Plant Pathology
448 Plant Science Hall
University of Nebraska
Lincoln, Nebraska 68583-0722
Telephone: (402) 472-2858

Table II. Sources of Local Information on Diagnosis
and Control of Turfgrass Diseases in the United States
and Canada (*Continued*)

NEVADA

Dr. Kathleen Kosta
Bureau of Plant Industry
Division of Agriculture
350 Capital Hill Avenue
Reno, Nevada 89502-2923
Telephone: (702) 688-1178

NEW HAMPSHIRE

Plant Diagnostic Laboratory
Department of Plant Biology
241 Spaulding Hall
University of New Hampshire
Durham, New Hampshire 03824
Telephone: (603) 862-3841

NEW JERSEY

Department of Plant Pathology
Box 231, Foran Hall
Cook College
Rutgers University
New Brunswick, NJ 08903
Telephone: (908) 932-9425, 932-9375

NEW MEXICO

Extension Plant Pathologist
Box 3AE; Plant Sciences
Cooperative Extension Service
New Mexico State University
Las Cruces, New Mexico 88003

Table II. Sources of Local Information on Diagnosis
and Control of Turfgrass Diseases in the United States
and Canada (*Continued*)

NEW YORK

Insect and Plant Disease Diagnostic Laboratory
Department of Plant Pathology
334 Plant Science Building
Cornell University
Ithaca, New York 14853-4203
Telephone: (607) 255-7850

NORTH CAROLINA

Plant Disease and Insect Clinic
Department of Plant Pathology
Room 1104 Williams Hall
Box 7616
North Carolina State University
Raleigh, North Carolina 27695-7616
Telephone: (919) 515-3619

NORTH DAKOTA

Plant Diagnostic Clinic
Department of Plant Pathology
P.O. Box 5012, Walster Hall
North Dakota State University
Fargo, North Dakota 58105
Telephone: (701) 231-7058

OHIO

Plant and Pest Diagnostic Clinic
Department of Plant Pathology
2021 Coffey Road
Ohio State University
Columbus, Ohio 43210-1087
Telephone: (614) 292-5006

Table II. Sources of Local Information on Diagnosis
and Control of Turfgrass Diseases in the United States
and Canada (*Continued*)

OKLAHOMA

Plant Disease Diagnostic Laboratory
Department of Plant Pathology
110 Noble Research Center
Oklahoma State University
Stillwater, Oklahoma 74078-9947
Telephone: (405) 744-9961

OREGON

Plant Disease Clinic
Extension Plant Pathology
Cordley Hall 1089
Oregon State University
Corvallis, Oregon 97331-2903
Telephone: (503) 737-3472

PENNSYLVANIA

Plant Disease Clinic
Department of Plant Pathology
220 Buckout Lab.
Pennsylvania State University
University Park, Pennsylvania 16802
Telephone: (814) 865-2204

RHODE ISLAND

For homeowners:
Cooperative Extension Education Center
University of Rhode Island
East Alumni Avenue
Kingston, Rhode Island 02881-0804
Telephone: (401) 792-2900

For professionally managed turf:
Department of Plant Sciences
Woodward Hall
University of Rhode Island
Kingston, Rhode Island 02881-0804
Telephone: (401) 792-5995

Table II. Sources of Local Information on Diagnosis and Control of Turfgrass Diseases in the United States and Canada (*Continued*)

SOUTH CAROLINA

Plant Problem Clinic
Cherry Road
Clemson University
Clemson, South Carolina 29634-0377
Telephone: (803) 656-2069

Pee Dee Research and Education Center
2200 Pocket Road
Florence, South Carolina 29506
Telephone: (803) 662-3526

SOUTH DAKOTA

Plant Disease Clinic
Department of Plant Science
Box 2109
South Dakota State University
Brookings, South Dakota 57007
Telephone: (605) 688-5123

TENNESSEE

Plant and Pest Diagnostic Center
University of Tennessee
P.O. Box 110019
Nashville, Tennessee 37222-0019

TEXAS

Texas Plant Disease Diagnostic Center
Room 101, L. F. Peterson Building
Texas A & M University
College Station, TX 77843-2132
Telephone: (409) 845-8033

Texas Agricultural Research and Extension Center
17360 Coit Road
Dallas, Texas 75252
Telephone: (972) 231-5362

Table II. Sources of Local Information on Diagnosis
and Control of Turfgrass Diseases in the United States
and Canada (Continued)

UTAH

Plant Pest Diagnostic Laboratory
Department of Biology
Utah State University
Logan, Utah 84322-5305
Telephone: (801) 797-2435

VERMONT

Plant Diagnostic Clinic
Department of Plant and Soil Science
Hills Building
University of Vermont
Burlington, Vermont 05405-0086
Telephone: (802) 656-0493

VIRGINIA

Plant Disease Clinic
106 Price Hall
Department of Plant Pathology, Physiology and Weed Science
Virginia Polytechnic Institute and State University
Blacksburg, Virginia 24061-0331
Telephone: (540) 231-6758

WASHINGTON

Serving eastern Washington:
Plant Diagnostic Clinic
WSU-Prosser-IAREC
Route 2 Box 2953-A
Prosser, Washington 99350-9687

Serving western Washington:
Plant Diagnostic Clinic
WSU-Puyallup Research and Extension Center
7612 Pioneer Way East
Puyallup, Washington 98371-4998

Table II. Sources of Local Information on Diagnosis
and Control of Turfgrass Diseases in the United States
and Canada (*Continued*)

WEST VIRGINIA

Plant Disease Diagnostic Clinic
401 Brooks Hall, P.O. Box 6057
Downtown Campus
West Virginia University
Morgantown, West Virginia 26506
Telephone: (304) 293-6023

WISCONSIN

Plant Pathogen Detection Clinic
Department of Plant Pathology
1630 Linden Drive
University of Wisconsin
Madison, Wisconsin 53706
Telephone: (608) 262-0928

WYOMING

Plant Disease Clinic
Department of Plant, Soil and Insect Sciences
P.O. Box 3354, University Station
University of Wyoming
Laramie, Wyoming 82071-3354
Telephone: (307) 766-3103

Canada

ALBERTA

Brooks Diagnostics LTD
Plant Diagnostic Laboratory
c/o Crop Diversification Centre-South
Brooks, Alberta T1R 1E6

Department of Land and Horticultural Sciences
Olds College
Olds, Alberta T4H 1R6
Telephone: (403) 556-4790

Table II. Sources of Local Information on Diagnosis
and Control of Turfgrass Diseases in the United States
and Canada (*Continued*)

BRITISH COLUMBIA

Plant Diagnostic Laboratory
British Columbia Ministry of Agriculture
and Fisheries
Abbotsford Agriculture Centre
1767 Angus Campbell Road
Abbotsford, British Columbia V3G 2M3

MANITOBA

Crop Diagnostic Centre
201-545 University Crescent
Agricultural Service Complex
Winnipeg, Manitoba R3T 5S6

NEW BRUNSWICK

Plant Diagnostic Laboratory
New Brunswick Department of Agriculture
and Rural Development
P.O. Box 6000
Fredericton, New Brunswick E3B 5H1

NOVA SCOTIA

Plant Diagnostic Laboratory
Nova Scotia Department of Agriculture
and Marketing
Kentville Research Station
Kentville, Nova Scotia B4N 1J5

Plant Diagnostic Laboratory
Department of Biology
Nova Scotia Agricultural College
Box 550
Troro, Nova Scotia B2N 5E3

Table II. Sources of Local Information on Diagnosis and Control of Turfgrass Diseases in the United States and Canada (*Continued*)

ONTARIO

Pest Diagnostic Laboratory
University of Guelph
P.O. Box 3650
95 Stone Road West
Guelph, Ontario N1H 8J7

Pest Diagnostic Clinic
Agriculture and Food Ser. Centre
P.O. Box 3650
95 Stone Road East, Zone 2
Guelph, Ontario N1H 8J7

Turf Extension Specialist
Ontario Ministry of Agriculture
Guelph Turfgrass Institute
Guelph, Ontario N1H 6H5

Department of Environmental Biology
University of Guelph
Guelph, Ontario N1H 6H5

PRINCE EDWARD ISLAND

Plant Health Services
Box 1600
Charlottetown, Prince Edward Island C1A 7N3
Telephone: (902) 368-5600

QUEBEC

Laboratoire de diagnostic
Le Service de Recherche en Phytotechnie de Quebec
2700, rue Einstein
Ste.-Foy, Quebec G1P 3W8